P9-AFV-122

Antimicrobial Food Additives

Springer

Berlin
Heidelberg
New York
Barcelona
Budapest
Hong Kong
London
Milan
Paris
Santa Clara
Singapore
Tokyo

R

Erich Lück · Martin Jager

Antimicrobial Food Additives

Characteristics · Uses · Effects

2nd revised and enlarged edition
translated from the German by S. F. Laichena

Springer

Dr. rer. nat. Martin Jager
Hoechst Aktiengesellschaft
Abteilung Lebensmitteltechnik
Postfach 80 03 20
D-65926 Frankfurt

Translation:
S. Laichena
15 Woodbury Park Road
Ealing
London W13 8DD
UK

Title of the German Edition:
E. Lück, M. Jager: Chemische Lebensmittelkonservierung, 3. Auflage, 1995

ISBN 3-540-61138-x 2. ed. Springer-Verlag Berlin Heidelberg New York
ISBN 3-540-10056-3 1. ed. Springer-Verlag Berlin Heidelberg New York

CIP data applied for

Die Deutsche Bibliothek – CIP-Einheitsaufnahme

Lück, Erich:
Antimicrobial food additives : characteristics, uses, effects /
Erich Lück ; Martin Jager. Transl. from the German by S. F.
Laichena. – 2., rev. and enl. ed. – Berlin ; Heidelberg ; New
York ; Barcelona ; Budapest ; Hong Kong ; London ; Milan ;
Paris ; Santa Clara ; Singapore ; Tokyo : Springer, 1997
Einheitssacht.: Chemische Lebensmittelkonservierung <engl.>
ISBN 3-540-61138-x
NE: Jager, Martin:

Typesetting: Fotosatz-Service Köhler OHG, Würzburg
Cover design: Konzept & Design, Ilvesheim
SPIN: 10506579   52/3020 – 5 4 3 2 1 0 – Printed on acid-free paper

For Ulrike, Lena and Jonas

Preface to the second English Edition

Since the publication of the first English translation of this book about 15 years ago, a great deal of new information on antimicrobial food additives has emerged in the fields of microbiology, toxicology and analysis. These aspects have been given prominence in this edition.

Despite a widespread and in most cases emotive aversion to food additives in general and preservatives in particular, the commercial use of preservatives is increasing rather than declining. There are many reasons for this: the growing popularity of convenience foods is certainly as important as changes in consumer awareness, against a background of continuous rises in the incidence of food infection and food poisoning.

The structure of the book remains unchanged. Like the previous edition it consists of two sections. The first section contains information applicable to preservatives in general. The second section describes the properties and applications of the various preservatives. The chapters are grouped into those on inorganic and organic preservatives respectively.

Each chapter has been thoroughly revised and updated. Preservatives that have become less important and those of little commercial relevance have been summarized fairly briefly in a chapter on "Other preservatives". This edition still focuses primarily on the commercial use of preservatives in the food sector. The book is aimed at those involved on the practical side in the food industry who have an interest in the scientific aspects of their work. Information on the regulatory status has deliberately been confined to the main points since a detailed and up-to-date description cannot be given against a background of constantly changing regulations.

Autumn 1996 Dr. Erich Lück, Dr. Martin Jager

Foreword to the first English Edition

Although the preservation of food by chemical techniques is such an important field of research in food science and also a major branch of the food industry, no monograph on the subject has hitherto existed in modern literature. Knowledge concerning this field is widely dispersed in a multiplicity of journals and handbooks. The foremost works of reference on food microbiology and food technology treat the preservation of food by chemical means only as a peripheral aspect.

This book first appeared as a German-language publication in the Federal Republik of Germany in 1977, since when it has sold widely throughout Europe. It reviews all aspects of food preservation by chemical techniques, the majority of which involve the use of chemical additives. This, the first English-language edition, is more than a straight translation from its Geman predecessor since the text has been updated in the light of new knowlegde aquired in the interim.

Essentially, this book is a collection of facts augmented by information drawn from literature but also contains much of the author's personal experience. It consists of two sections:

1. a general section dealing with aspects of importance to all preservatives and
2. special chapters on the properties and uses of individual preservatives.

The sequence of chapters in the second edition of the book follows the usual system employed in inorganic and organic chemistry. A separate chapter is devoted to each substance that is, or used to be, of major practical importance. Preservatives which have attained a minor degree of importance at some time are then summarized in a further single chapter. This book also deals with those disinfectants and gases which may still be regarded as preservatives in the broader sense of the term because they are used for determination of foods. In other respects the text confines itself strictly to food preservation; only brief reference is made to food analysis since good books on the analysis of preservatives in foods already exist.

The book is aimed mainly at the practical man in the food industry with an interest in the scientific aspects of his work. The scientific principles of food preservatives are all explained in sufficient detail for a clear understanding of the way they are used. In addition, the book is intended as a reference work in which prominence is given to the major interrelationships in its field of reference. In the chapters dealing with the individual preservatives a deliberate attempt has been made to provide systematic description enabling the student or other reader to obain a rapid overall picture. Finally, in deciding on the book's layout and scope it has been born in mind that a book of this type can also serve as a source of information for government authorities, medical practitioners, nutritional scientists and, not least, the educated layman. It is hoped the book will help bring objectivity to all emotionally charged discussion on the use of preservatives in foods and return such discussion to a scientific level.

The author would like to record his sincere gratitude to Grant F. Edwards, Manager of the Translation Department of Hoechst UK, for his careful and conscientious translation work. Thanks are also due to the publishers, notably Dr. Boschke, for their critical comments, of which due account has been taken in both the German and English editions, as well as for their promptitude in completing the task of publication.

Spring 1980 Dr. Erich Lück

Table of Contents

General Considerations

The Individual Preservatives

General Considerations

Chapter 1

Aim and Development of Food Preservation

All foods are prone to spoilage and must therefore be consumed immediately or, where this is not possible or desirable, preserved so that they can be stored.

1.1 Food Spoilage: Definition and Controlling Factors

The quality of food can be adversely affected by physical, chemical, biochemical and microbiological processes. This review, however, will confine itself to food spoilage caused by microorganisms and to the substances that reduce the rate of spoilage or actually prevent it altogether.

The biological processes involved in food spoilage due to microorganisms cannot take place unless certain extrinsic conditions are met:

1) The presence of spoilage microorganisms:
   Spoilage cannot occur unless microorganisms are (still) present on or in the food.
2) Nutrients for the microorganisms need to be bioavailable:
   Unless the microorganisms have nutrients available which they can utilize, their life cycle cannot commence.
3) Conditions favorable to life must exist in respect of:
   a) Temperature
   b) Water activity
   c) Presence or absence of oxygen
   d) Redox potential
   e) pH value
   Unless these environmental conditions are favorable or at least adequate, growth will be slowed down or not occur at all.
4) Sufficiently long food storage time:
   If the food is consumed before any undesired growth of microorganisms occurs, measures against intrinsically possible microbiological spoilage become superfluous.
5) The vital activity of the microorganisms must cause a deterioration in the quality of the food:
   Unless microorganisms on the food become recognizable and/or excrete harmful or otherwise undesirable metabolic products, the term "spoilage" is inapposite.

Not every microbiological change in a foodstuff is regarded as spoilage of the food. The expression "food spoilage" is confined to what, by convention and in normal parlance, would be termed undesirable. Hence, the intentional fermentation of grape juice by yeast with a view to producing wine is not food spoilage, although the undesired fermentation of the self-same grape juice would indeed class as spoilage if the intention were to retain the grape juice as such. Vinegar results from the spoilage of wine, hence the French word 'vinaigre' (vinegar); the intentional production of vinegar from wine by the same vinegar bacteria is not classed as spoilage, however. Occasionally, the scale of the reaction and the course it takes is the determining factor as to whether a microbiological change in the food is termed spoilage or not. This principle is exemplified by the microbiological process occurring in the ripening of cheese. Depending on its extent, the microbiological process is termed ripening (desired) or spoilage (undesired). The transition from one to the other is gradual, and sometimes the point at which spoilage begins is not clear-cut. Since most countries operate a ban on the marketing of spoilt food, this uncertainty may also have legal consequences. Without elaborating the point, it should nevertheless be mentioned at this juncture that certain microorganism strains are an essential requirement for the production of certain foods, e.g. bread, yoghurt and wine.

1.2 Definition of and Processes for Food Preservation

Food preservation in the broad sense of the term refers to all measures taken against any spoilage of food. In its narrower sense, however, food preservation connotes the processes directed against food spoilage due to microbial action. Since ancient times two fundamentally different types of method have been used, namely physical and chemical methods. In recent times biological methods of food preservation (see Sect. 30.25 and 30.26) have become increasingly important.

The microorganisms need not necessarily be killed. It is much more important to produce conditions that merely prevent the microorganisms from causing food spoilage.

Physical preservation methods, which will not be described in this book, are notable in that the food is subjected to a physical measure which counters microbial growth. The best-known physical methods are those of sterilization and pasteurization (heat treatment), cooling and freezing (refrigeration), drying (dehydration), and irradiation.

The chemical methods are characterized by the addition of a chemically more or less readily definable substance that inhibits the development of microorganisms or kills them. These substances are known as preservatives. A distinction is drawn between preservatives in the broader sense of the term and those in the narrower sense. Examples of the former group are common salt and vinegar, whilst examples of the latter are sorbic acid and sulfur dioxide. The essential difference between the groups is in their applied concentration. The preservatives in the broader sense are used in concentrations upwards of 0.5 – 1%, whereas those in the narrower sense can be employed in concentrations of 0.5% or less because their antimicrobial action is more powerful.

In the biological methods, high purity, harmless microorganism cultures, which have an inhibitory effect on undesirable spoilage microorganisms, are added to the foods. These are known as "protective cultures".

1.3 The Need for Food Preservation

Cereals, fruit and other foods of plant origin are available fresh only during a brief harvest period. Consequently, since ancient times it has been necessary to keep supplies of harvested food edible over relatively long spells. Rainy periods and droughts, which have occurred everywhere in the world and continue to do so, have likewise necessitated proper preservation of foods. One of the earliest references is in the Bible (Genesis, chapter 41, verses 34–36), where Pharaoh commanded Joseph to store one-fifth of the harvest in granaries during the seven fat years in order to provide supplies for the seven lean years. This was the first recorded instance of food preservation by means of protective gas, since the cereal was protected from spoilage as a result of the carbon dioxide it produced by respiration.

In these instances the need for food preservation was engendered entirely by natural factors, but later the factors of civilization were also involved. Man is living increasingly in towns and cities, where it is no longer possible to plant or harvest one's own food on any large scale. In all the industrialized countries, a declining number of people are responsible for obtaining and producing food for a growing number of other people. Such a development is possible only if the foods have adequate keeping power.

Recently, moreover, there has been a change of living habits and requirements in many countries, inasmuch as people now wish to enjoy foods and specialties from distant lands. Many foods whose keeping properties are subject to highly exacting requirements have become branded articles. These are products that can be made available only by a suitable preservation technique specific to the product in question.

Although food preservation has now reached a high standard, at least in the developed countries, the quantities of food still spoiled are astonishing. According to estimates, no less than 20% of foods produced never reach the consumer's table because of attack beforehand by rodents, insects or microorganisms. In less-developed regions of the world the percentage of food lost to human consumption is far higher.

Whereas food used to be preserved for commercial reasons alone, toxicological findings have recently become additional reason for preservation. Since the nineteen-sixties it has been known that some molds form aflatoxins and other mycotoxins, and that these molds can enter foods. If the growth of molds is inhibited, for example by the use of preservatives, toxin formation can be reduced (Lück 1981). As a form of preventive medicine therefore, the use of a toxicologically harmless preservative sometimes poses a smaller risk than failure to use one.

1.4 History of Chemical Food Preservation

At the time when man was still a gatherer and hunter who lived, in the truest sense of the term "from hand to mouth", he had no need of food preservation. It was unnecessary for food to stay fresh for any lengthy period since nature constantly provided fresh supplies in adequate quantities. Not until the neolithic revolution some ten thousand years ago, in the New Stone Age, did man begin to adopt a settled life-style. Then, instead of gathering and hunting, he took to tilling the soil and keeping animals. This forced him increasingly to lay in stores. Owing to his lack of specialist knowledge, man confined himself in those early days to simply stockpiling his food. This he collected much as a squirrel collects its nuts; then protected it against theft by his fellow-men and the vagaries of wind and weather. In doing this he was forced to sacrifice much of the food's nutritive and organoleptic quality.

Initially, the preservation technique involve drying and salting. The diet was governed accordingly. From careful reading of old accounts about the food of seafarers or the winter diet of country-folk, who were partially or completely dependent on preserved foods, at least in the temperate zones of the world, it will be found that the diet was dominated on the one hand by cereals and flour and on the other hand by dried, salted or pickled meat or salted/unsalted dry fish. Thus a fourteenth-century Parisian merchant advised his customers to soak a twelve-year old stockfish overnight and then beat it with a blacksmith's hammer until it was tender. In a number of countries it was normal practice to bake bread only two or three times a year, after which it would be dried out and eaten in the ensuing months in the form of a softened mush. Obviously, such a diet was monotonous and ailments due to deficiency diseases were not uncommon, especially since little or nothing was known about the influences of the preservation methods on the constituents of the foods. In time, the list of preservatives used grew to include alcohol, smoke, sulfur dioxide and a number of organic acids, such as acetic and lactic acid; then, for the next two thousand years these constituted the total range available.

Food preservation changed with the commencement of industrialization. The need for food preservation increased rapidly, and people became more fastidious. No longer were they satisfied with the preservatives mentioned above, since these produced a radical alteration in the structure and properties of the foods they preserved.

The advances made in chemistry were also utilized in preservation techniques. Thought began to be given to the principles underlying the preservatives employed up to that time. In his investigations into smoke, K Reichenbach discovered, in pyroligneous acid and the tar of beechwood, an oil that he named creosote owing to its property of preserving meat. Reichenbach wrote in glowing terms of its preserving action, although already at that time he stated that the substance involved health risks (Strahlmann 1974). Creosote was probably used on only a limited scale for sensory reasons; yet it is the only preservative to be described in any detail in a well-known book on food chemistry dating from 1848, apart from salt (whose use was correctly described as indirect drying), drying, heat treatment, lactic acid fermentation, sugars, alcohol, vinegar and smoke (Knapp 1848).

For rather more than a century now, increased efforts have been made to preserve foods not simply by any feasible means but to do so in a way that leaves their frequently sensitive constituents unimpaired, as well as safeguarding their nutrient value and flavor against unfavourable influences. In the first flush of excitement at the discovery of substances with an antiseptic action for medicinal purposes, "chemicals" such as hydrofluoric acid, fluorides, chlorates and the like were used in

Table 1. Historical development of chemical food preservation (Strahlmann 1974)

Prehistoric times	Common salt, smoke
Ancient Egypt	Vinegar, oil, honey
Ancient Rome	Sulfur dioxide for stabilizing wine (?)
Before 1400	Invention of pickling by Beukels
1775	Borax recommended by Höfer
1810	Sulfur dioxide recommended for preserving meat
1833	Creosote recommended for preserving meat by Reichenbach
1858	Antimicrobial action of boric acid discovered by Jaques
1859	Sorbic acid isolated from rowan berry oil by Hofmann
1865	Antimicrobial action of formic acid discovered by Jodin
1874	Antimicrobial action of salicylic acid discovered by Kolbe and Thiersch
1875	Antimicrobial action of benzoic acid discovered by Fleck
1907	Formaldehyde and hydrogen peroxide recommended for milk preservation by von Behring
1908	Benzoic acid permitted for use in foods in the USA
1913	Antimicrobial action of *p*-chlorobenzoic acid discovered by Margolius
1923	Antimicrobial action of *p*-hydroxybenzoic acid esters discovered by Sabalitschka
1938	Propionic acid recommended for the preservation of baked goods by Hoffman, Dalby and Schweitzer
1939	Antimicrobial action of sorbic acid discovered by Müller and, independently of this, by Gooding in 1940
1947	Antimicrobial action of dehydroacetic acid discovered by Coleman and Wolf
1950 onwards	Worldwide revision of approvals for new preservatives
1954	Industrial-scale production of sorbic acid commences
1956	Antimicrobial action of diethyl pyrocarbonate discovered by Bernhard, Thoma and Genth
1980 onwards	Increasing use of protective gases

food preservation. In no sense was this done unscrupulously as a means of profiteering or with intent to deceive; it was merely due to ignorance of the products' potential harm, since toxicological investigations were unknown. People simply believed that the small quantities of a substance just sufficient to act as a preservative could scarcely be harmful; so at first they were undiscriminating in their choice of substances. Hence, the introduction of salicylic acid and boric acid to food preservation about a hundred years ago should be regarded as an advance, although nowadays both these preservatives are considered outmoded. At the end of the second half of the nineteenth century, formic acid joined the preservatives. The beginning of the twentieth century saw the first use of benzoic acid, which is still widely used in food preservation. As an aromatic compound benzoic acid, like salicylic acid, was regarded unfavorably from the outset. Effective derivatives were therefore sought and, as a result, *p*-chlorobenzoic acid and the esters of *p*-hydroxybenzoic acid were discovered. This was followed in the late nineteen-thirties by the salts of propionic acid, and in the post-war years by sorbic acid and its salts. The introduction of sorbic acid is largely the result of the worldwide toxicological re-evaluation, which commenced around 1950, of food ingredients in general and preservatives in particular. As an unsaturated fatty acid, sorbic acid is the most thoroughly researched and harmless of all preservatives in widespread use.

Over the last 15–20 years the trend towards consuming fresh foods has grown substantially. Logistical measures implemented by the food industry have progressively shortened the transportation times between production sites and the consumer. Refrigeration techniques, also employed during the transport of food, have shown vast improvements in the industrialized nations.

Another current trend is the increasing discussion concerning "harmless" additives for food preservation purposes, i.e. the use of antimicrobial substances produced from plants and microorganisms. Many laymen regard such products as less suspicious than others merely because of their natural origin.

It is worth noting that in the many articles attacking the use of additives it is the preservatives that receive the least criticism, on the whole. This is because even critics realize that in certain cases preservatives do indeed protect consumers from harmful effects on their health. In future, therefore, antimicrobial food additives will retain their importance in certain areas.

1.5 Literature

Knapp FC (1848) Die Nahrungsmittel in ihren chemischen und technischen Beziehungen. Vieweg, Braunschweig, p. 101–109

Lück E (1981) Schutzmaßnahmen gegen Lebensmittelverderb durch Schimmelpilze. In Reiß J: Mykotoxine in Lebensmitteln. Gustav Fischer, Stuttgart – New York, p. 437–457

Strahlmann B (1974) Entdeckungsgeschichte antimikrobieller Konservierungsstoffe für Lebensmittel. Mitt Geb Lebensmittelunters Hyg 65, 96–130

Chapter 2

Analysis of Preservatives

Owing to food law requirements and technological considerations, it is essential that preservatives in a foodstuff should be capable of determination by qualitative analysis and that quantitative measurement of the preservative should also be possible.

2.1 Qualitative Determination

Since the individual food preservatives cannot be grouped in a single class of substances with a standard chemical definition, a generally applicable method of sample preparation and analysis is scarcely feasible. It is, however, possible to use non-specific microbiological tests to determine whether a preservative is present or not. For this purpose the foodstuff to be investigated, possibly after appropriate dilution, is inoculated with defined strains of microorganism, on which the preservative is known to have an antimicrobial action. A study is then made over a particular period of time to ascertain whether the microorganisms multiply. Many microorganisms, especially yeasts, release carbon dioxide, the evolution of which can be measured. Known as the "fermentation test", this method used to be of some importance, but is rarely used today because it is not sufficiently specific.

Defined strains of microorganism can be obtained from the "DSM" (German microorganism collection) in Braunschweig, Germany or the "American Type Culture Collection" (ATCC, 12301 Parklawn Drive, Rockville, MD 20852, USA).

Wet assay and instrumental methods of determination for virtually all food preservatives have been described in the literature. The principles of each of these methods are described in the chapters on the individual substances themselves. The purpose of these descriptions is to present the range of methods available for detecting each preservative, examining the advantages and drawbacks of each method. This should enable the analytical chemist to select the appropriate method. However, full experimental details are not given; these can be found in the literature references.

2.2 Quantitative Determination

Foodstuffs have an extremely complex composition and consist of many individual constituents which may interfere with chemical detection of preservatives. For this reason and sometimes also in order to increase their concentration, the preservatives to be determined nearly always have to be isolated from the food as a first step. To do this it is necessary to use solid-liquid or liquid-liquid extraction, steam distillation or special methods specific to individual cases. Possibly after further purification, the extract can then be made available for actual analysis. This method is generally known as sample preparation.

Quantitative determination of the preservatives can be successfully carried out with gas, thin-layer or liquid chromatography, wet chemical processes, colorimetry, photometry and other methods, depending on the substances to be tested. In view of the increasingly widespread use of high-performance liquid chromatography (HPLC), the focus will be on this method in the analytical section of each chapter describing an individual substance and on the relevant detection techniques.

2.3 Purity Requirements

Only preservatives of special and standardized purity can be used as food additives. Hence, in most modern regulations on the acceptance of preservatives, there are increasing numbers of special purity requirements for the individual substances (see 2.4 General literature). In general they are concerned chiefly with the content of toxicologically relevant heavy metals and specific impurities deriving from the process of synthesis.

2.4 General Literature

Analysis

Bundesamt für das Gesundheitswesen, Abteilung Vollzug Lebensmittelrecht (Publisher) Schweizerisches Lebensmittelbuch (5th edition), Ring book III, Chapter 44 "Konservierungsstoffe für Lebensmittel", Bern, Eidgenössische Drucksachen- und Materialzentrale, loose-leaf collection (1992 onwards)

Horowitz E (1984) Official methods of analysis of the Association of Official Analytical Chemists. 14th edition. Association of Official Analytical Chemists, Washington

King RD (1978) Developments in food analysis techniques. Applied Science Publishers, London

Kommission des Bundesgesundheitsamtes zur Durchführung des § 35. LMBG (Publisher). Amtliche Sammlung von Untersuchungsverfahren nach § 35. LMBG, Beuth, Berlin, loose-leaf collection (1980 onwards)

Purity requirements

Council of Europe (Editor) (1989) European Pharmacopoeia (2), European Treaty Series No. 50, Sainte-Ruffine: Maisonneuve

FAO and JECFA (Editor) (1990) Food and nutrition paper (FNP), Specifications for identity and purity of certain food additives (continuing series), FAO/WHO, Rome, issue no. 49

FAO and JECFA (Editor) (1992) Compendium of food additive specifications, Addendum 1, FAO/WHO, Rome, issue no. 52

Glandorf KK, Kuhnert P, Lück E (1991) Handbuch Lebensmittelzusatzstoffe, Behr, Hamburg, chapter CV, loose-leaf collection (1990 onwards)

National Research Council (US) Food and Nutrition Board (1981), Food Chemicals Codex: (FCC), 3rd edition. National Academy Press inc. supplements

Verordnung über das Inverkehrbringen von Zusatzstoffen und einzelnen wie Zusatzstoffe verwendeten Stoffe (Zusatzstoff-Verkehrsverordnung) of July 10, 1984

CHAPTER 3

Health Considerations

3.1 General Considerations

Food has always contained substances which, in certain concentrations and under particular conditions, may threaten or damage health. These may be naturally occurring food ingredients, such as goitrogenous substances in rape and other types of vegetables or cyanogenic glycosides in cassava. Alternatively such substances may either form in plants themselves as a reaction to exogenous noxae (so-called phytoalexins such as solanine or chaconin in potatoes) or, after microbial attack, may help contaminate food with mycotoxins or other toxins. In view of the high toxicological potential of most mycotoxins, this particular type of contamination is a problem that cannot be ignored (Goto 1990, Jelinek et al. 1989). Preservatives, when used appropriately and correctly, can prevent the formation of mycotoxins in many foods and, as a result, greatly improve food safety.

Auxiliaries and additives, including preservatives, used to be employed in foods without prior testing. The absence of adverse effects on the health after the consumption of food treated with preservatives was accepted as adequate proof that they were harmless. This meant that people were "testing" the substances added to the food as well as the food itself.

Deriving from that time, when "chemicals" were used in food preservation without safeguards, there is still a discernible but diminishing aversion in some circles to food preservation generally, even today. Since the beginning of this century and especially since the nineteen-fifties, the situation has changed radically with the development of toxicology as a specialist discipline. Auxiliaries and additives are now permitted and employed for food purposes only if the toxicological tests conducted upon them in accordance with the level of scientific knowledge at the time give no reason for believing they are in any way harmful, or even that they might be so. In this connection it should be pointed out that in the case of additives the legislators refer to the "principle that they are prohibited unless expressly permitted". This means that all additives are prohibited and only those expressly approved (in terms of permissible maximum quantity and field of use) may actually be used. Today the harmlessness to health of auxiliaries and additives for foods is better researched than that of some foods and food ingredients. It is interesting to note that the natural occurrence of toxic ingredients in foods arouses much less public attention than the presence of food additives.

In tests to determine the toxicological properties of auxiliaries and additives, the assumption is made that a dose-effect relationship exists (a mathematically quantifiable relationship between dosage, duration of effect and the extent of the effect) as well as the fact that there is a limit below which the substance has no effect in the organism ("no-effect level"). This limit must be determined. Unlike the general pharmacological aim of determining when effects occur, the objective here is to discover when they fail to occur. As a scientific approach, this is rather unusual. It does not, however, apply to carcinogenic substances with a genotoxic activity mechanism, since the immense biological effect (covalent modification of the DNA) of such substances makes it impossible to give a no-effect level (Hemminki 1993).

As long ago as 1538, in the third of his seven "Carinthian Defences" (Epistola dedicorata St. Veit), Paracelsus formulated the principle that substances must be able to exist in doses too small to produce an acute effect. To quote Paracelsus himself (in translation):

What exiſteth, but that it be poiſonſome?
All thingſ be poiſonſome
and none there be without poiſon.
Nought but the doſe maketh
a thing poiſonleſſ.
An example thereof:
any mete or drynke
in ſurfete
be poiſonſome
aſ the effecte ſheweth.
I do admit
that poiſonſ alſo be poiſonſome.

Paracelsus did not imply by this that a substance becomes a poison only above a certain dosage but that a poison ceases to have an acutely toxic effect below a certain dosage.

The scope of the required tests and the way in which they are conducted depends on the use the substance being tested is specified for or expected to perform. Nowadays animal experiments are preceded by a number of tests known as in-vitro short-time assays. These are used to assess parameters such as genotoxicity and mutagenicity (e.g. micronucleus assay, HGPRT test, Ames mutagenicity test, see Sects. 3.4 and 3.5). Aspects such as "irritant effect" or "penetration of the skin" can also be assessed by in-vitro test systems that are already available (e.g. HET-CAM "chorionallantois" test), as can acute cytotoxicity (e.g. neutral red uptake inhibition test). Even if every known in-vitro test is used in assessing a substance, this cannot entirely obviate the need for animal experiments. Preliminary in-vitro testing may simply reduce the number of animal experiments (Spielmann 1989). It is also necessary in each individual case to clarify whether such in-vitro results can be applied firstly to animals and secondly to man and whether they can be validated ("What is the significance of a positive result in an in-vitro assay for man and how far are the results comparable?").

For obvious reasons, the basic knowledge required for a toxicological assessment of a substance is derived from the results of animal experiments. Most of the animals involved are small, short-lived creatures such as mice and rats, and, depending on the parameters being studied, other rodents, dogs, monkeys and, in special instances, other animals, too. The animals used are always bred specially for the purpose and kept under defined conditions prior to the experiment. For studies of certain aspects that are more pharmacological in nature, genetic engineering is used to breed experimental animal species which serve as excellent animal models.

In advanced stages of trials, as in the case of pharmaceuticals, the biochemical behavior and metabolic reactions in man are investigated in human volunteers under medical supervision. This greatly reduces the risk involved in transferring experience gained in animal experiments to man.

The following criteria are nowadays regarded as especially important for assessing the harmlessness of an auxiliary or additive:

1) Acute toxicity,
2) Metabolic investigations and toxicokinetics,
3) Genotoxicity/mutagenicity,
4) Reproductive toxicity including fertility toxicity and teratogenicity,
5) Subchronic toxicity,
6) Chronic toxicity,
7) Carcinogenicity.

This list of criteria forms the basis for the flow chart in Fig. 1. Frequently experiments are combined, e.g. chronic toxicity experiments and carcinogenicity tests. During metabolic experiments, tests are carried out to discover whether the substance being studied accumulates in the organism. It is also necessary to check whether the biotransformation (absorption, distribution, metabolism and elimination) of a test substance proceeds along similar lines.

The present-day practice whereby all toxicological investigations are conducted at institutes specializing in the relevant sectors ensures that suitable types of animal are used, that the animals are properly kept and fed, and that the substance under test is administered properly and in the correct dosage. Appropriate test guidelines can be obtained from the OECD, for example. Finally it is vital that results should be properly interpreted.

Before the toxicological investigation of an auxiliary or additive can begin, its identity must be determined. The substance being tested must be chemically and physically identical to the substance actually used subsequently in practice. Therefore the first step is to draw up precise specifications of the test substance. Besides the usual chemical data these must state the quantity of pure substance and any impurities the test substance may contain. It is generally easier to draw up specifications for synthetic or natural substances with a precisely defined chemical specification than to do so for natural substances with a complex composition. In the case of the precisely defined substances the possibility of by-products occurring can be deduced from the method of synthesis, where this is known. These by-products are, however, usually removed from the substances to be tested by the purification operations normally employed in chemical technology today, such as recrystallization and distillation. Although a residual quantity of impurities may in some cases be very important, the toxicological evaluation should not overlook the question of dosage in this context either. Auxiliaries and additives are used in foods in small to minute quantities. Impurities in auxiliaries and additives in the ppm to ppb range must therefore be viewed differently from equivalent amounts of contaminants in foods themselves.

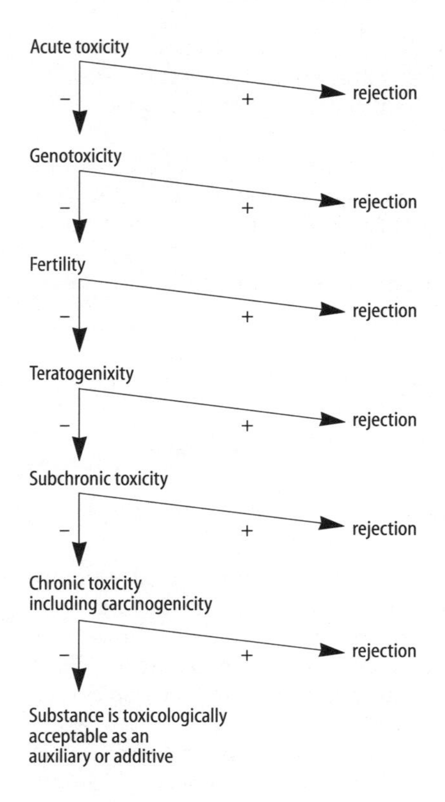

Fig. 1. Stepwise assessment of the effect of auxiliaries and additives on health

It is regarded as fundamentally desirable for food additives not to display any pharmacological effects in the concentrations used. In principle, this requirement also includes preservatives. It has to be borne in mind, however, that preservatives also have an effect on pathogenic microorganisms, mainly owing to their antimicrobial action, and could thus theoretically also be used to combat them, e.g. fungal infections of the skin. The effect of the substances used as food preservatives in controlling pathogenic microorganisms is, however, much too weak for them to be realistically considered for such an application. To prevent the development of resistance, substances used in medical practice are not permitted for use as food additives. This applies in particular to a number of antibiotics.

3.2 Acute Toxicity

Acute toxicity, expressed in the form of the LD_{50}, is merely an approximate means of measuring the toxicological properties of a substance after single-dose administration (Zbinden and Flury-Roversi 1981). The LD_{50} is the dose at which 50% of the animals in an experimental group are expected to die. It is governed by a number of external factors, e.g. the species of animal, the age, the weight, the sex and the conditions under which the animal is kept. All these factors must be standardized.

The LD_{50} is stated in milligrams of test substance per kg body weight of the experimental animal. The higher the value, the lower the acute toxicity of the substance.

For gaseous substances an LC_{50} (= median lethal concentration) is given in mg/l air. The LD_{50} and LC_{50} are different from the ED_{50} (effective dose 50%), i.e. the quantity of a substance at which a particular effect is produced in 50% of the animals.

To determine the LD_{50}, the test substance is administered in various doses to groups of animals, each comprising five males and five females. For auxiliaries and additives used in foods, only peroral administration is important. The amount being investigated is administered as a single dose through the esophagus. If the quantity is too great because the substances are of very low toxicity, it can also be administered in several smaller portions, though on the same day. The animals are observed for at least one to two weeks.

First, the LD_{50} forms the basis for classifying a substance in comparison with similar substances of known acute toxicity. This it does by indicating the dose at which effects on the animal can be expected. It also indicates the profile of toxicological effects and thus acts as a basis for determining the range of doses used in subchronic and chronic feeding tests. It is also possible in some cases, after autopsies on the animals, to discover which organs were particularly affected by the test substance. These observations also provide guidance on planning and evaluating subsequent toxicity investigations. Finally, the acute toxicity determined yields information on expected risks to man in the event of accidents or misuse, or after handling of the relevant substance at work.

Accordingly, the LD_{50} can be used to place test substances in the following poison classes:

Poison class 1 = LD_{50}	< 5 mg/kg body weight
Poison class 2 = LD_{50}	5–49 mg/kg body weight
Poison class 3 = LD_{50}	50–499 mg/kg body weight
Poison class 4 = LD_{50}	500–4999 mg/kg body weight
Poison class 5 = LD_{50}	> 5000 mg/kg body weight

The following table shows that all substances with a preservative action (other than nitrites = poison class 3) belong in poison classes 4 and 5.

Table 2. LD_{50} (oral) of various preservatives

	LD_{50} (mg/kg body weight)	Animal species
Benzoic acid	3000	rat
Dehydroacetic acid	1000	rat
Biphenyl	3300	rat
Ethanol	9500	mouse
Formic acid	1200	mouse
p-Hydroxybenzoic acid ester	6000–8000	mouse
Nitrates	6000	rat
Nitrites	100– 200	rat
o-Phenylphenol	3000	rat
Pimaricin	1500	rat
Propionic acid	4000	rat
Salicylic acid	1100	rabbit
Sodium chloride	3750	rat
Sorbic acid	10000	rat
Sucrose	30000	rat

3.3 Metabolic Investigations and Toxicokinetics

Data concerning the absorption, breakdown and, if applicable, conversion of the test substance in the organism is no less important to the overall toxicological evaluation of a substance than information on the body organs attacked. Investigations providing such data are also carried out initially on animals.

Investigations involving animals which metabolize the test substance in the same or at least a similar way to man are particularly informative. Human volunteers are brought into the investigations at a later stage in the experiment and under medical supervision.

Prior to the start of the animal experiments, a general overview is obtained of the probable behavior of the test substance in the organism. Acid, alkali and enzyme hydrolyses are tested in vitro and/or on isolated organs, cells or cell organelles.

Monitoring the sequence of stages that a substance passes through in the body as it is broken down involves administering it in normal and raised concentrations and then studying the conversion and excretion of the substance itself and/or the formation of any metabolites in bodily fluids, organs and tissues. By administering raised doses it is possible to detect whether the metabolism may be dose-related (formation of different metabolites if the usual metabolic route is "overloaded"). This study reveals the site, i.e. the organ or tissue, where the substance under test undergoes biochemical changes and identifies any dose-time relationships that exist. To achieve this, the substance being investigated is synthesized, in most cases by using the radioactive carbon isotope ^{14}C, and administered to the experimental animal.

To study several sections of a molecule separately, it is possible to apply double labels to the target molecule (using ^{3}H and ^{14}C, for example) and detect ^{3}H and ^{14}C activity separately in the organs and body fluids of the animal. If the initial intention is merely to investigate the overall distribution of radioactivity in the experimental animals, this can be achieved by autoradiography. In autoradiography an animal, e.g. a mouse or rat, is given a dose of test substance labelled with radioactive isotopes and then killed, after which the whole animal is deep-frozen in liquid nitrogen. Median sections are placed on a photographic plate, where the distribution of overall radioactivity (test substance and its metabolites) in the animal body can then be detected very clearly.

It is not possible to differentiate between an intact test substance and metabolites by autoradiography. This can be done after processing each organ, preparing extracts from it and separating these extracts by HPLC. An HPLC radioactivity detector is then used to differentiate between test substance and metabolites. To clarify the structure of the metabolites the usual methods can be employed (using the UV spectrum, IR spectrum, ^{1}H and ^{13}C spectrum) but mass spectroscopy is the most commonly used technique. All metabolic studies form an important basis for subsequent subchronic and, in particular, chronic feeding trials.

There are several possible alternatives for the metabolic behavior of a preservative:

Insoluble substances are generally excreted through the intestines in unchanged form. Biological effects outside the gastro-intestinal tract are not likely to occur, nor is the formation of metabolites to be expected. It is, of course, necessary to ensure that the test substance is not converted by intestinal bacteria into breakdown products that can be absorbed. It is only in the case of very fine-particle substances that there is a possibility of pinocytosis, phagocytosis or persorption, i.e. the direct transfer of the substance from the intestine to the blood stream. There are no examples of preservatives that come within this group.

A second group contains substances which are absorbed from the gastrointestinal tract but not chemically changed. Such substances can then be excreted in unchanged form through the urine, without the formation of toxicologically relevant metabolites. With these substances it is also necessary to check whether intestinal bacteria cause chemical changes. There are no preservatives that come within this group of substances either.

A third group comprises substances which, although absorbed from the gastro-intestinal tract, are not excreted by the organism until after biochemical breakdown or biotransformation has occurred. They are, for example, oxidized in phase I reactions, and then bonded in phase II reactions by glucuronization, sulfation, phosphatization or some other process and thereby made hydrophilic and eliminable. One example of such a substance is benzoic acid, which in humans is bonded to glycine and excreted in the form of hippuric acid through the urine. In the case of such substances that can be metabolized it is important that biotransformation occurs relatively quickly and that no metabolites accumulate in the body.

A fourth group consists of substances which, like those described previously, are absorbed and may be chemically changed. In this case, however, excretion of these substances or of any metabolites occurs relatively slowly. Such substances can accumulate in the organism, which is undesirable. Examples of preservatives of this type are boric acid and salicylic acid.

The last group contains substances which the body utilizes to generate calories after absorption in the same way as it does with food itself. Such substances undergo the same biotransformation as food constituents, whether they are fats, proteins, carbohydrates or other ingredients. Examples of such substances are propionic acid and sorbic acid.

3.4 Genotoxicity

The terms genotoxicity, genetic toxicity and mutagenicity cover all the damaging effects of a substance on genetic material. Spontaneous mutations can occur in any living organism, but they are rectified extremely quickly by special "repair systems" (for example, O^6-alkyl-guanine transferase). A distinction can be drawn between gene, chromosomal and genomic mutations. They are caused by bonding of the test substance to the DNA, inhibition of repair systems or other mechanisms.

Gene mutations are typically changes within a gene. In the simplest case, mutations can be caused by an altered base pair:

1) transition = replacement of one purine base by another, or one pyrimidine base by another;
2) transversion = replacement of a purine base by a pyrimidine base or vice versa;
3) addition = addition of one or more base pairs;
4) deletion = loss of one or more base pairs.

Addition and deletion mutations are also known as frame-shift mutations.

Chromosomal mutations, unlike gene mutations, can be recognized under an optical microscope from changes in the chromosome structure. Substances which induce chromosome aberrations are known as clastogenes. The most important types of chromosome aberrations are:

1) deletion,
2) duplication,
3) inversion,
4) translocation.

Ring chromosomes are another example of chromosome aberration.

Genomic mutations take two forms: aneuploidy and polyploidy. In complete aneuploidy, entire chromosomes are affected. Hence, if a chromosome appears once in the diploid set this is known as monosomy and if it occurs three times, as trisomy. In polyploidy it is the whole chromosome set that occurs several times, and not just individual chromosomes.

Mutagenicity can be investigated by means of in-vitro tests with microorganisms, e. g. bacteria, yeasts, fungi and other cell cultures, as well as in-vivo tests with plants, insects, especially *Drosophila melanogaster*, small rodents, especially mice and, finally, lymphocyte cultures from human subjects. Since the various methods are based on different sites attacked by the test substances in the genetic substrate, mutagenicity testing needs to consist of a whole "battery" of tests.

The instructions for carrying out a number of these tests (Ames mutation test, HGPRT test, cell transformation test, chromosomal aberration test, UDS test, micronucleus test and the dominant lethal test) have been published in the EC Gazette or by the OECD. However, the results obtained in these tests require careful interpretation, since both false positive and false negative results can occur.

The Ames test is by far the best-known. It is based on the fact that certain strains of *Salmonella typhimurium* which have mutated and consequently lost their ability to synthesize histidine, revert to their original type under the influence of mutagens and are then able to grow on histidine-free nutrients. When preservatives are tested in the Ames test, problems can arise since preservatives may have an inhibitory effect on *Salmonella typhimurium*. Different strains of *Salmonella typhimurium* (e.g. TA 98, TA 100, TA 1535) vary in their sensitivity to certain types of mutation (e.g. frame-shift mutations) and so the strain of *Salmonella typhimurium* selected can itself give an indication of the type of induced mutation that may be obtained. Another of the tests frequently carried out is the micronucleus test (see Sect. 3.13 General literature).

Mutagenicity is a very important factor in the toxicological evaluation of auxiliaries and additives for the food sector, because close relationships nearly always exist between mutagenicity and carcinogenicity, at least for genotoxic substances.

3.5 Reproductive Toxicity

Studies of reproductive toxicity involve testing male and female fertility and the general ability to reproduce, as well as testing for teratogenic properties and the effect on peri- and postnatal development.

The term "teratogenicity" connotes the induction of malformations in the embryo or fetus. In principle, any substance with a teratogenic effect would be unacceptable as an auxiliary or additive in the food sector. The naturally occurring food ingredient alcohol, for example, has considerable teratogenic potential and the consequent embryopathy associated with alcohol is described as a serious toxicological problem.

To test for teratogenicity, the substance being investigated is administered to pregnant animals in the critical phases of prenatal development. Among the aspects determined are: the number of implantations, early or late absorption, living and dead fetuses, position and distribution of fetuses in the horns of the uterus, weight of the litter, as well as malformations of the skeleton and organs.

At different times during prenatal development different organs display varying degrees of sensitivity to teratogenic noxae, and so in studies of teratogenesis the timing of exposure is an important factor. The teratogenic effect of thalidomide ("Contergan catastrophe"), of alcohol and of diethylstilboestrol (DES) have all been very well documented, although the last-mentioned constitutes a special case as it involves the transplacental induction of tumors. There are also indications of teratogenic potential in folic acid antagonists (aminopterin) and synthetic progestins (Hodgson and Levi 1992).

3.6 Subacute Toxicity

Subacute toxicity tests, which normally last 28 days, form a link between acute toxicity investigations (single administration with a one- to two-week observation period) and subchronic toxicity tests (daily administration of the test substance for 90 days, see Sect. 3.7). In subacute toxicity tests, multiple applications (usually three) within a short period (28 days) enable any cumulative toxic effects of a test substance or metabolites to be recorded. During the tests it is also possible to detect whether the organism adapts to the test substance and whether the substance targets certain organs in preference. This can be more precisely determined by measuring the activity of various liver enzymes.

3.7 Subchronic Toxicity

Subchronic toxicity investigations involve feeding a substance to an experimental animal over a period of 90 days to 6 months, this period corresponding to approximately 10% of the life span of the selected animal. For the tests it is customary to use a rodent, e.g. rats, mice or hamsters and a non-rodent, e.g. dogs or pigs. To detect any gender-related differences in effect, male and female animals are always included in the tests in equal numbers. The number of animals used must be sufficient to allow statistical evaluations. The tests include a number of control animals which are kept under the same conditions, but not fed with the test substance. The substance under test is administered in several (at least three) doses. The highest dose should be in a quantity at which toxic effects can be expected. This makes it possible to detect which organs the test substance attacks, and to focus special attention on these organs in the long-term experiment.

Essentially, the subchronic toxicity experiment comprises:

1) tests on the live animal;
2) tests on the animal after it has been killed at the end of the experiment.

While the animal is alive the experiment involves investigating and observing external phenomena, such as behavior, mobility and changes in body weight, as well as feed and water consumption. The urine and feces are regularly examined to check their appearance and composition, and to identify the presence of unphysiological substances or other deviations from their normal composition. In addition, clinico-chemical investigations are carried out on the blood and blood serum to obtain information about the performance of important organ functions. At the end of the experiment the animals are killed. Important internal organs are weighed and then examined both macroscopically and histologically. The liver and kidneys are especially important because of their central metabolic and excretory functions. They often react to the administration of high doses of certain substances by becoming slightly and reversibly enlarged but without undergoing histological changes. In such cases, the organ enlargements can be explained as biological reactions to a stress situation.

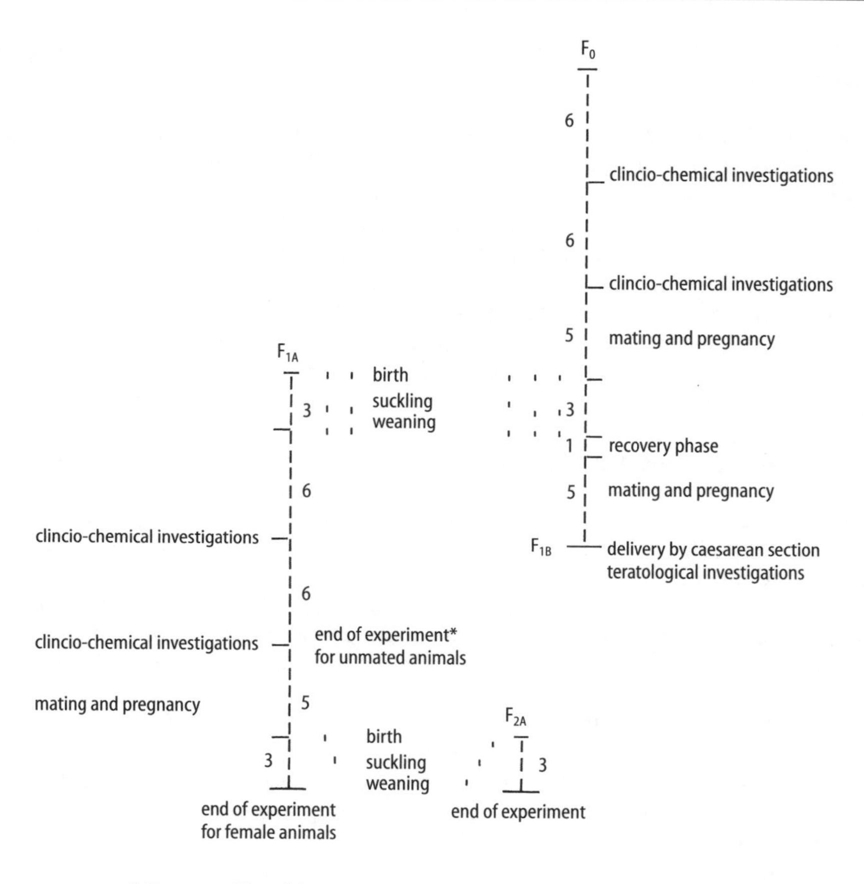

Fig. 2. Sequence of stages in a subchronic toxicity test on the rat combined with reproduction toxicology investigations

The doses of test substance selected for subchronic toxicity experiments should be such that, although toxic effects are clearly discernible, the experimental animals survive. Besides the test aims already mentioned, namely to investigate "adaption", "accumulation", "reversibility of damage" and "organs attacked", the other purposes of the subchronic toxicity test are to determine the dose (by acting as a range-finding study) and the application method for chronic toxicity studies (see Sect. 3.8).

3.8 Chronic Toxicity

The term "chronic toxicity" connotes the sum effect detectable after administering a substance to rats or mice over 2 years or more. This most closely reflects the actual situation when auxiliaries and additives are consumed in foods. Small quantities of such substances may also be consumed over very long periods, sometimes throughout an individual's life. Unlike the subchronic toxicity test, where substances are administered in relatively high doses, the emphasis in this test is on administering substances over long periods, so that other damage can be detected, such as any carcinogenic effects or phenomena dependent on the susceptibility of certain tissues at a particular age. The chronic toxicity test is therefore regarded as the main method of evaluating the potential risk of an additive when used in food.

One crucial requirement in the chronic feeding experiment is that a sufficiently large number of animals is used. This enables the results to be evaluated by statistical methods, limits the wide range of scatter in the levels of biological activity and shows up effects that rarely occur. In other respects the same principles apply as in the subchronic toxicity experiment (see Sect. 3.7).

The test substance is administered to the animals in relatively high doses. The doses are selected on the basis of the results of the preceding subchronic feeding experiments. The intended applied concentration in the food and the possible consumption quantity are also relevant.

The largest dose is generally the highest one at which no effect, such as organ damage, was caused in the subchronic toxicity experiment. It is not advisable to administer doses exceeding 5%, relative to the feed, as this might cause stress-related conditions, osmotic effects or other influences and hence unspecific phenomena producing results possibly unrelated to the substance. Excessive doses of substances may disrupt the balance of calories in the diet or affect the utilization of nutrients., unduly high applied concentrations of a test substance can produce organoleptic changes in the feed, which may sometimes be rejected by the experimental animals as a result. This can have an unfavorable effect on the animals' feed intake and growth.

The observations in the chronic toxicity experiment concentrate initially on the animals' growth and behavior, externally detectable functions of their body organs and enzyme reactions. The experimental animals are either killed at a specified point in time, namely the predetermined end of the experiment, or they are kept until they die of natural causes. An autopsy is carried out on every animal. The histopathological investigations initially focus on the group of animals given high doses. The following organs and tissues are examined for macroscopic lesions and tumors (including local lymph nodes): lymph nodes, mammary glands, salivary glands, femur or vertebra, including bone marrow, hypophysis, trachea, lungs, heart, thyroid gland, esophagus, stomach, small intestine, colon, liver, gall-bladder, pancreas, spleen, kidneys, suprarenal glands, bladder, prostate gland, testis, ovaries, uterus, brain, eyes and spinal cord. If there are any unusual features in the distribution of a substance in the organism, then it may also be necessary to examine other organs and tissues.

At the end of the experiment the results must be properly interpreted. The prime concern is to assess whether they can be transferred to man and to calculate any risk. Food additives have to satisfy higher safety requirements today than do pharmaceuticals. Certain side effects can be accepted more readily in pharmaceuticals, if their use counteracts some greater risk to the health, than in food additives, which must be "free of side effects" in the human organism.

After all the evaluations have been completed, a dose is ascertained at which no effects can be detected. This "no observed effect level" (NOEL), which represents the highest dose at which *no* toxicological effects occur, is used as the basis for setting the acceptable daily intake (ADI) (see Sect. 3.11). If the tolerable dose in the long-term feeding experiment is divided by the applied concentration in certain foods, this gives the "current safety margin".

Table 3 lists the current safety margins of various preservatives. The higher the value for the current safety margin, the less harmful a substance is considered to be. One striking feature is that long-familiar substances not even defined as preservatives in international legislation have the lowest current safety margin. This disregards the fact that additives expressly permitted as preservatives are allowed in only a few foods in legally stipulated maximum quantities, whereas substances such as common salt and sugar are present in many foods for reasons other than preservation and in much higher concentrations than preservatives in the narrow sense.

Table 3. Current safety margin of various preservatives

	Chronic tolerability (% in feed)	Applied concentration in directly margin consumed	Current safety foods
Acetic acid	10	1%	10
Benzoic acid	1	0.1%	10
Common salt	1	2%	0.5
Biphenyl	0.1	0.005%	20
Ethanol	4	up to 30%	abt. 0.13
Formic acid	0.2	0.3%	0.7
Nitrite	0.02	0.01%	2
PHB ester	1	0.05%	20
Propionic acid	3	0.3%	10
Sorbic acid	5	0.1%	50
Sugar	about 60	up to 60%	abt. 1
Sulfur dioxide	0.20	0.02%	10

3.9 Carcinogenicity

Since auxiliaries and additives for the food sector are ingested with food over long periods of time, even by children and young persons, the possibility of such substances participating in the various stages of carcinogenesis must be examined extremely carefully. It is assumed today that the development of tumors is a long-term process involving many stages, which can be subdivided into the phases of exposure, initiation, promotion, conversion and progression (multistep concept) (Harris 1991, Sugimura 1992).

In the context of this concept, it should be noted that substances can act as initiators in carcinogenesis (e.g. genotoxic agents) and at the same time as promoters, encouraging the formation of tumors (e.g. phorbolester or ethanol). Only the interaction of initiators and promoters, followed by the steps of conversion (change of a precarcinogenic lesion into a malignant tumor) and progression (growth and metastasis) leads to the full development of cancer.

The initiators of carcinogenesis are, in most cases, genotoxic substances, i.e. substances which are able to produce covalent modification of DNA. Besides genotoxic carcinogens, there are also non-genotoxic (epigenetic) carcinogens (Lijinsky 1990, Green 1992) such as saccharin, limonene, or BHA, whose effect, however, has so far been demonstrated only on very specific animal species. In addition it is always necessary to check in the case of genotoxic agents whether a linear or non-linear dose-response relationship exists. In formaldehyde, for example, the dose-response relationship is non-linear and so a threshold limit can be stated even for formaldehyde-induced carcinogenesis. The substances most relevant to the development of tumors, however, are genotoxic agents and so DNA adducts, which are evidence of covalent modification of DNA, are particularly important nowadays in evaluating the carcinogenic potential of test substances. For Aflatoxin B_1 it is possible to use both immunological methods and HPLC to detect the formation of DNA adducts in man. The use of such methods for risk assessment is known as "molecular dosimetry" (Hemminki 1993, Kadlubar 1992, Marshall 1993).

Since lengthy induction periods may pass before tumors occur, any carcinogenic effect is studied by feeding a substance to animals over their entire life span, commencing at the earliest possible age. The animals most commonly employed in these experiments are mice and rats because their life span is relatively short. Control studies with feed that does not contain the test substance are used in order to ascertain the spontaneous tumor rates in the relevant animal species and thus avoid drawing false conclusions. In tests for carcinogenicity it is extremely important that the number of animals kept under observation should be sufficiently large; this principle applies both to experimental animals and to control animals. In general, two species of animal are included in such studies.

Carcinogenicity studies are protracted and expensive. For this reason and not least to protect the animals from unnecessary experimentation, considerable efforts are being made to find suitable short-time tests. No such tests have so far been sufficiently validated, however, and those found do not completely eliminate the need for animal experiments. Mutagenicity tests can be carried out first (see Sect. 3.4). However they do not necessarily obviate the risk that some substances not themselves carcinogenic (so-called precarcinogens) may be capable of transformation (activation) in the body into reactive metabolites. Only these "ultimate carcinogens" can then, for example, undergo covalent modification with the DNA and act as initiators. To test for the possibility of such a reaction, the radioactively labelled test substance is applied to rats. After a few hours' to days' reaction time, the DNA is isolated in the body organs being studied, chiefly the liver, and its radioactivity is determined. This indicates the quantity of bound test substance and is known as the CBI (covalent binding index). Carcinogenic and non-carcinogenic substances differ in the extent of their covalent binding by a factor of 1000–100000 (Lutz and Schlatter 1979). There are also many techniques available today for determining covalent DNA modification (Hemminki 1993).

In testing preservatives, studies are basically confined to administering the test substance in the feed. The occurrence of sarcomas after injecting solutions of a substance should not be regarded as proof of carcinogenicity so long as the sarcomas remain localized at the site of the injection and provided that no other indications of a carcinogenic effect exist. This is because tissue damage and inflammation caused by chronic irritation may occur if repeated injections are given and these conditions may themselves give rise to sarcomas although totally unrelated to the action of the test substance.

In the case of well-known substances already employed in food chemistry for some time, epidemiological studies for assessing the risk of carcinogenicity are of some use, although their informative value is disputed because of the associated uncertainties (e.g. observation is only retrospective) and the difficulty in interpreting them. For substances going through the approval process, epidemiological studies are, in any case, impossible.

One major problem in evaluating possible carcinogenic effects of food additives is that of the threshold limit, or rather the dose at which no undesired effects are caused. Arguments supporting what some authors postulate to be the "irreversibility" of carcinogenic effects are based on dose-response studies, single-dose and multi-generation experiments, as well as on the concept of somatic mutation as the first steps in carcinogenesis with subsequent transmittance of induced defects during cell replication. The problem of extrapolating the results of animal experiments using high doses to the situation in humans has not yet been solved satisfactorily, because it must be assumed that exposure in humans is low, at least as far as the dose is concerned. Possible practical consequences include zero tolerance, acceptable threshold limits with low risk and safety factors.

It is known from animal experiments that many substances in high doses have a carcinogenic effect, whereas low doses produce no tumors at all. Formaldehyde is such a substance.

There are also many substances which occur naturally in small quantities in foods, but are carcinogenic in high concentrations. Moreover, animal experiments involving high concentrations suggest that even substances, such as certain foods or alcohol, which are consumed in large quantities may promote if not initiate tumors of the esophagus, rectum and mammary glands, for example (Blot 1992, Muir and McKinney 1992). This has led to a fraught situation in the USA, where the so-called Delaney anticancer clause is legally binding. It states that a substance may not be permitted for use in the food sector, even below a threshold level, if it has a carcinogenic effect in humans or animals in any applied concentration, however high. The Delaney anticancer clause, usually referred to simply as the Delaney clause, applies only to additives and not to foods themselves.

3.10 Allergenic Effect

For scientific purposes, the symptoms described colloquially as "food allergy" must be identified either as genuine food allergies (immune reactions produced by antigen-antibodies) or "food intolerant reactions" (not produced immunologically). Contrary to popular belief, the commonest causes of such reactions are not additives in general or preservatives in particular but naturally occurring food ingredients such as certain proteins in nuts, cows' milk or certain cereals (Aulepp and Vieths 1992, Thiel 1991).

Genuine food allergies are what are known as immediate response allergies (type I reactions). As in the case of hay fever, for example, these occur as a response to the increased formation of specific antibodies (immunoglobulins E and IgE) against epitopes of the allergenizing agent. As the immunological reaction proceeds, the antibodies bond together in twos, this being followed by degranulation of the mast cells, accompanied by a release of mediators such as histamine or bradykinine. These mediators are ultimately responsible for the symptoms of genuine food allergies. When such reactions tend to run in families, the condition is described as "atopy". These allergy types should be distinguished from type IV reactions, which involve T cells but not antibodies, and are also described as "late types" (for example, allergy to nickel in fashion jewellery). As initially stated, a distinction must be drawn between genuine allergic reactions and food intolerant or pseudoallergic reactions (PAR). Food intolerant reactions can be caused by both innate and acquired enzyme defects; examples are lactose intolerance and alcohol intolerance in many Asians. The symptoms of intolerant reactions, like those of pseudoallergic reactions, are often difficult to distinguish from genuine allergies (Vieths et al. 1993 and 1994).

Descriptions can be found of intolerant reactions and allergies to a number of additives, such as food dyes, antioxidants, technical enzymes and preservatives. It should be noted, however, that although certain substances with a preservative action, such as salicylic acid, have long ceased to be used as additives, they may still cause allergies because they occur naturally. *p*-Hydroxybenzoic acid and its esters, as well as benzoic acid, are still employed as preservatives even today, and so reactions to them can be attributed both to their use as additives and to their natural occurrence. Since the use of such additives must be declared on food labels, allergy sufferers can avoid foods containing them.

Among the preservatives used today, those most often described as allergens are sulfites, especially for asthmatics (Belchi Hernandez 1993, Nagy et al. 1995), benzoic acid (Jacobsen 1992, Michils et al. 1991) and parabens (Jacobsen 1992), whereas propionic acid and sorbic acid have no allergenic effect (Rosenhall 1982).

3.11 Acceptable Daily Intake (ADI)

The most reliable method of testing the harmlessness of a preservative would undoubtedly be actual experience with human subjects. However, for ethical reasons trials of this type are possible only at a very advanced stage in any investigation, and then only on a limited scale. Hence, researchers have to deduce how a preservative will behave in humans from the results of animal experiments.

Using the results of toxicological studies as a basis, joint expert committees of the FAO/WHO (JECFA) and EU scientific food committee (SCF) establish what is known as the acceptable daily intake (ADI). The ADI is the quantity of an auxiliary or additive, measured in milligrams, which can be ingested daily per kilogram body weight without misgivings throughout the entire life span. The ADI acts as a guide. Quantities slightly higher than the ADI are acceptable (Truhaut 1991, Kroes et al. 1993).

The ADI is the quotient obtained from the highest dose of an auxiliary or additive that still causes no toxic effects in the chronic toxicity experiment (NOEL see Sect. 3.8), divided by the safety factor, which is normally 100. It can be raised if there are special reasons for doing so, or lowered if the substances involved are constituents of human food or if the way in which the relevant substance is broken down is similar to that of food ingredients (Renwick 1991 and 1993). The safety factor of 100 is ten times the factor 10 and allows a margin for various imponderables and assessment risks:

1) In toxicological studies of food additives, which are conducted mainly on animals for obvious reasons, there is a definite risk – albeit a small one – in applying the results of animal experiments to humans. The possibility that a certain animal may be especially sensitive or particularly insensitive to a certain food additive cannot always be totally excluded, but the risk of this is covered by the factor of 10.
2) When allowance is made for difference in body weight a number of animals have a higher food intake than man. The rat, for instance, eats seven or eight times as much as man, relative to body weight.
3) Food frequently contains several additives at the same time. Although synergistic effects in toxicological behavior are rare, they must nevertheless be allowed for.
4) Among those who ingest food additives are persons having special metabolic behavior patterns, e.g. children, invalids and the elderly, as well as expectant and nursing mothers. These additional risks are included expressly in the ADI concept. These circumstances also are allowed for by a factor of 10, thus producing a factor of 100 when multiplied by the factor of 10 referred to in point 1.
5) The ADI applies to the entire life span and also covers the interests of groups in need of special protection. In addition it allows for variations in personal eating habits.

In cases where the expert committees consider that the toxicology of a substance has not been sufficiently clarified, a "temporary ADI" is sometimes laid down. The term "ADI not limited" is used to describe substances with especially low toxicity or those representing essential constituents of foods or normal metabolic products in the human organism.

Should toxicological misgivings arise concerning a substance already specified, it is possible to change the permanent ADI to a temporary ADI while awaiting the results of further studies. Thus in February 1994 the SCF changed the ADI values

Table 4. ADI values of preservatives (JECFA)

	acceptable daily intake (mg/kg body weight)
Acetic acid, including its salts	no limit
Benzoic acid, including its salts	0–5[1]
Biphenyl (Diphenyl)	0–0.05
Formic acid, including its salts	0–3[1]
Hexamethylene tetramine	0–0.15
p-Hydroxybenzoic acid ethyl ester	0–10[1]
p-Hydroxybenzoic acid methyl ester	0–10[1]
p-Hydroxybenzoic acid propyl ester	0–10[1]
Hydrogen peroxide	not determined
Lactic acid, including its salts	no limit
Lysozyme	acceptable
Natamycin (Pimaricin)	0–0.3
Nisin	0–33000 (units)
o-Phenylphenol	0–0.2
Propionic acid, including its salts	no limit
Sodium diacetate	0–15
Sodium and potassium nitrate	0–5
Sodium and potassium nitrite	0–0.2 (not for baby food)
Sorbic acid including its salts	0–25[1]
Sulfur dioxide, including sulfites	0–0.7[1]

[1] Group ADI (in some cases calculated as acid).

for benzoic acid and parabens, which had previously been permanently set at 0–5 mg and 0–10 mg/kg body weight respectively, to temporary ADI values, and plans to consult again in 1996 when further studies are completed. The FAO/WHO has not yet adopted this change in classification and so, although the two committees have identical ADI values, the importance they attach to the figures differs. There may even be differences in the ADI values set by the two committees (JECFA and SCF).

For preservatives, the values listed in table 4 currently apply.

The NOEL (no observed effects level) of the test substance, on which the ADI is based, relates to body weight. In contrast, the feeding experiment is always based on the content of test substance in the feed. Hence, two different measuring systems are employed. Each of these can be converted to the other as shown in Table 5. The conversion table takes account of the body weight of the experimental animals and the feed quantities normally consumed.

It is difficult to apply the ADI direct to food regulations, which generally stipulate the maximum content of a preservative in a food, whereas the ADI relates to the quantity of the relevant substance consumed per day. Nevertheless, the ADI is increasingly being used as a guide by legislators when approving additives and especially when setting maximum quantities. In view of the built-in safety factor in the ADI, it is possible to set values for a preservative right up to its full ADI especially since exceeding the ADI briefly is not thought to pose any risk.

The actual consumption of food additives is nearly always far below the ADI. Careful calculations for preservatives in the Federal Republic of Germany show that the population consume only some 1–10% of the ADI (Lück and Remmert 1976). However there are some exceptions, e.g. sulfur dioxide.

The ADI must be distinguished from the permissible level (PL), tolerable daily intake (known in German as the DTA) and the provisional tolerable weekly intake (PTWI). The PL is stated in ppm in foods and is obtained by multiplying the ADI by body weight and dividing the result by the proportion of average food consumption that the preserved food constitutes. It therefore indicates how high a concentration of additive may be present in various foods, assuming average consumption levels, without the ADI being exceeded. The tolerable daily intake is calculated by the Federal German Office of Health for pesticides for which there are no ADI values (JECFA) (Anon. 1993, Anon. 1994). PTWI values (FAO/WHO) are mainly calculated for heavy metals and residues of substances with a pharmacological action.

Table 5 Conversion of toxicity data from one form to another

Animal species	Average weight	Average feed consumption per day	1 ppm in feed corresponds to mg/kg body weight/day	1 mg/kg body weight/day corresponds to ppm in feed
Mouse	0.02 kg	3 g	0.150	7
Young rat	0.1 kg	10 g	0.100	10
Adult rat	0.4 kg	20 g	0.050	20
Guinea pig	0.75 kg	30 g	0.040	25
Rabbit	2.0 kg	60 g	0.030	33
Dog	10.0 kg	250 g	0.025	40

3.12 Preservative Mixtures

It is common practice to employ mixtures of various preservatives, which might conceivably display toxicological behavior different from that of the individual substances. In fact, no such difference occurs in the LD_{50} or the subchronic toxicity when between two and 20 percent of the LD_{50} is administered in the feed over a period of six months. The tests included the following products, some in the form of their sodium salts: dehydroacetic acid, sorbic acid, benzoic acid, *p*-hydroxybenzoic acid ethyl, propyl and butyl esters, salicylic acid, propionic acid and furyl furamide. Only a combination of benzoic acid and sulfite behaves rather more unfavorably in the chronic test (Shtenberg and Ignat'ev 1970).

Mixtures of sorbic acid with nitrite in high doses are no more toxic than equivalent quantities of nitrite alone. When a mixture of the 13 food additives most commonly used in Japan was administered over a period of up to 12 months, slight toxic effects were produced only for a quantity 10 times the tolerable daily intake and more pronounced toxic effects for an amount 100 times the tolerable daily intake; the food additives subjected to these tests included the preservatives sodium benzoate, potassium sorbate, sodium propionate, *p*-hydroxybenzoic acid butyl ester, biphenyl and the sodium salt of dehydroacetic acid.

3.13 General Literature

Bakterientest: Rückmutationsversuch mit *Salmonella typhimurium* (Gen)mutationstest (= Ames test). Offic J Europ Commun 27, 143–145 (19.9.1984, L 251)
Chromosomenaberrationstest in vivo (in vivo mammalian bone marrow cytogenetic tests chromosomal analysis): method published by the OECD (OECD Guideline 475) dated 4.4.1984
Chromosomenaberrationstest in vitro: Versuch zum Nachweis von Chromosomenschäden in Säugerzellen in Kultur, Offic J Europ Commun 27, 131–133 (19.9.1984, L 251)
DNS-Schädigung und -Reparatur (Unscheduled DNA synthesis = UDS) in Säugerzellen in vitro, Offic J Europ Commun, EWG Directive 79/831 Anlage 5 Teil B (March 1987)
In vitro-Säugerzell-Gen-Mutationstest: Offic J Europ Commun, EWG Directive 79/831 (March 1987)
In vitro-Zelltransformationstest: EWG Directive 67/548 (30.5.1988)
Mikrokerntest: Versuch zum Nachweis von Chromosomenschäden oder einer Schädigung des Mitoseapparates in vivo. Offic J Europ Commun 27, 137–139 (19.9.1984, L 251)
Säuger in vivo-Dominant-Lethal test. Offic J Europ Commun, EWG Direktive 87302, 76–78, 30.5.1988
Classen H-G, Elias P, Hammes W (1987) Toxikologisch-hygienische Beurteilung von Lebensmittelinhalts- und -zusatzstoffen sowie bedenklicher Verunreinigungen. Paul Parey Verlag, Berlin, p. 15–60
Eisenbrand G, Metzler M (1994) Toxikologie für Chemiker. Thieme Verlag, Stuttgart
Fahrig R (1993) Mutationsforschung und genetische Toxikologie. Wissenschaftliche Buchgesellschaft, Darmstadt
Marquardt H, Schäfer S (1994) Lehrbuch der Toxikologie. BI-Wissenschaftsverlag, Mannheim
Marquis J (1989) A guide to general toxicology, 2nd edn. Karger, Basel
Müller L, Miltenberger HG, Marquardt H, Madle S (1993) Die Bedeutung von in vitro Zelltransformationstests für die Routineprüfung. Bioforum 16, 69–76
Parke D (1993) Food, nutrition and chemical toxicity. Smith Gordon, London
Richardson M (1993) Reproductive toxicology. VCH Verlagsgesellschaft, Weinheim
Schettler G, Schmähl D, Klenner T (1991) Risk assessment in chemical carcinogenesis. Springer, Berlin Heidelberg New York
Schuhmacher G-H, Fanhänel J, Persaud T (1992) Teratologie. Gustav Fischer Verlag, Stuttgart
Shibamoto T, Bjeldanes L (1993) Introduction to food toxicology. Academic Press, Orlando
Thiel C (1992) Lebensmittelallergien und -intoleranzreaktionen. In: Ernährungsbericht 1992, Deutsche Gesellschaft für Ernährung e.V., Druckerei Henrich, Frankfurt
Timbrell J (1989) Introduction to toxicology. Taylor and Francis, London
Vainio H, Magee P (1992) Mechanisms of carcinogenesis and risk identification. IARC, Lyon
Walker A, Rolls B (1992) Nutrition and the consumer – issues in nutrition and toxicology 1. Elsevier Science Publishers, Essex
Walker R, Quattrucci E (1988) Nutritional and toxicological aspects of food processing. Taylor and Francis, London.
Würgler FE (1991) Genetische Endpunkte. Bioforum 14, 376–380

3.14 Specialized Literature

Anon. (1991) LD_{50} test (Akute Toxizitätsprüfung OECD Guideline 401) wird durch "fixed dose procedure" ersetzt. Alternative Tierexp Nr. 16, volume 3, 71
Anon. (1993) Festsetzung einer duldbaren täglichen Aufnahme (DTA-Wert, BGA, für Rückstände von Pflanzenschutzmittel-Wirkstoffen im Rahmen des Zulassungsverfahrens). Bundesgesundheitsbl 36, 247–249
Anon. (1994): ADI und DTA-Werte für Pflanzenschutzmittelwirkstoffe, edition 4 (issued on 5.1.1994), Bundesgesundheitsbl 37, 182–184
Aulepp H, Vieths S (1992) Probleme mit Nahrungsmittelallergien. Dtsch Lebensm Rundsch 88, 171–179
Belchi Hernandez J (1993) Sulfite induced urticaria. Annals of Allergy 71, 230–232
Blot W (1992) Alcohol and cancer. Cancer Res (Suppl) 52, 2119s–2123s
Goto T (1990) Mycotoxins: Current situation. Food Rev Intern 6, 265–290
Green S (1992) Nuclear receptors and chemical carcinogenesis. Trends in pharmaceutical science (TIPS) 13, 251–255
Harris C (1991) Chemical and physical carcinogenesis: advances and perspectives for the 90th Cancer Research (Suppl) 51, 5023s–5044s
Hemminki K (1993) DNA Adducts, mutations and cancer. Carcinogenesis 14, 207–212
Hodgson E, Levi P (1992) A text book of modern toxicology. Elsevier, New York, 159–168 and 256–260
Jacobson D (1992) Adverse reactions to benzoates and parabens. In: Food Allergies, Adverse Reactions to Foods and Food Additives, Blackwell Scientific Publication, Boston
Jelinik C, Poland A, Wood G (1989) Worldwide occurrence of mycotoxins in foods and feeds – an update. J Assoc Off Anal Chem 72, 223–230
Kadlubar F (1992) Detection of human DNA-carcinogen adducts. Nature 360, 189
Kroes R, Munro I, Poulsen E (1993) Workshop on the scientific evaluation of the safety factors for the acceptable daily intake (ADI): Editorial summary. Food Add Contam 10, 269–273
Lijinski W (1990) Non-genotoxic environmental carcinogens. J Environmental Science Health C8 (1), 45–87
Lück E, Remmert KH (1976) ADI-Wert und Lebensmittelrecht. Gedanken über Einflüsse und Auswirkungen, dargestellt anhand der Gesetzgebung über Lebensmittelkonservierungsstoffe in der Bundesrepublik Deutschland. Z Ges Lebensmittelrecht 3, 115–143
Lutz W, Schlatter C (1979) In vivo covalent binding of chemicals to DNA as a short time test for carcinogenicity. Arch Toxicol Suppl 2, 411–415
Marshall E (1993) Toxicology goes molecular. Science 259, 1394–1398
Michils A, Vandermoten G, Duchateau J, Yernault J-C (1991) Anaphylaxis with sodium benzoate. Lancet 337, 1424–1425
Muir C, McKinney P (1992) Cancer of the oesophagus: A global overview. European J Cancer Prevention 1, 259–264
Nagy S, Teuber S, Loscutoff S, Murphy P (1995) Clustered outbreak of adverse reactions to a salsa containing high levels of sulfites. J Food Protect 58, 95–97
Renwick A (1991) Safety factors and establishment of acceptable daily intakes. Food Add Contam 8, 135–150
Renwick A (1993) Data-derivate safety factors for the evaluation of food additives and environmental contaminants. Food Add. Contam 10, 275–305
Rosenhall L (1982) Evaluation of the intolerance of analgesics, preservatives and food colorants with challenge tests. Eur J Respir Dis 63, 410–419
Shtenberg A, Ignat'ev A (1970) Toxicological evaluation of some combinations of food preservatives. Food Cosmet Toxicol 8, 369–380
Spielmann H (1989) Die zentrale Erfassungs- und Bewertungsstelle für Ersatz- und Ergänzungsmethoden zum Tierversuch. Bundesgesundheitsbl 32, 360–363

Sugimura T (1992) Multistep carcinogenesis: A 1992 perspective. Science 258, 603–607
Thiel C (1991) Lebensmittelallergien und -intoleranzreaktionen. Z Ernährungswiss 30, 158–173
Truhaut R (1991) The concept of the acceptable daily intake: an historic review. Food Add Contam 8, 151–162
Vieths S, Schöning B, Aulepp H, Baltes W (1993) Identifizierung kreuzreagierender Allergene in Pollen und pflanzlichen Lebensmitteln. Lebensmittelchemie 47, 49–53
Vieths S, Fischer K, Dehne L, Aulepp H, Wollenberg H, Bögel K (1994) Versteckte Allergene in Lebensmitteln. Bundesgesundheitsbl 37, 51–60
Zbinden G, Flury-Roversi M (1981) Significance of the LD_{50} test for the toxicological evaluation of chemical substances. Arch Toxicol 47, 77–99

CHAPTER 4

The Legal Situation Relating to Food

4.1 Historical Development in Former Times

Early food legislation was concerned primarily with the control of weights and measures as well as other serious falsifications relating to foodstuffs, although the spoilage of food had long been a matter of public interest. The first decrees relating to food preservation were directed against the use of excess sulfur in wine. This practice was roundly condemned at the Diets held at Lindau as early as 1497 and at Freiburg in Breisgau in 1498 after Sebastian Brandt had mentioned problems with wine in his "Narrenschiff" (Ship of Fools) in 1494:

Much meddling there is now with wine,
One tries deception all the time.
Saltpeter, sulfur, bones of dead
Potash, milk, mustard, herbs ill-bred
Through bungs are pushed into the kegs.
The pregnant women drink these dregs,
That they bear children premature,
A wretched sight one can't endure ...

In the last century the regulations on preservatives became more concrete when the aim was to eliminate abuse (Strahlmann 1970). At the Tenth International Hygiene Congress in Paris in 1900, demands were made for a total ban on food preservation with chemicals. The legislative bodies, however, did not pursue this suggestion, which in its own way was just as excessive as the efforts of many producers and users of preservatives. It was appreciated that the use of preservatives within well-defined and sensible limits is justifiable and that without preservation with chemicals it is actually impossible to provide large groups of the population with an efficient and reliable food supply.

4.2 Recent Efforts at International Level

Since the nineteen-fifties there have been perceptible signs of increased efforts to harmonize food law internationally. The increasing movement of goods across national frontiers is only one of the reasons for this. Since 1954 the FAO and WHO have set down many bases for national and international regulations in meetings of joint committees of experts and within the framework of the Codex Alimentarius Commissions.

Until well into the twentieth century, food law regulations in many countries were confined to prohibiting the use of substances harmful to the health. Particular substances were actually named in some cases. This method, which failed to provide the consumer with adequate protection in the long term, has now generally been replaced by the system of so-called positive lists, by which, in principle, all substances are prohibited unless allowed by special regulations ("prohibited unless expressly permitted"). In general, further stipulations specify which substances may be used with which foods and in which maximum concentrations. In this connection, it is a general principle to allow preservatives only for use with those foods for which their application is sensible from the viewpoint of their effect. The permissible maximum quantities are determined by technical requirements with due allowance for a reasonable safety margin.

Stipulations for preservatives are frequently contained in special preservative regulations, as in the UK and formerly in the Federal Republic of Germany. In some countries preservatives are covered by the regulations on individual food products, as in France. Recently there has been an increasing tendency for preservatives to be regulated together with other food additives. This has been common practice for some time in Scandinavia, Switzerland and the Federal Republic of Germany, and is also due to be applied throughout the EU. The EU global directive on additives contains very extensive positive lists stating which preservatives may be used in which foods, the permissible maximum quantities and, where appropriate, the conditions of use.

Table 6. List of preservatives permitted for use in foods within the European Union

E 200	Sorbic acid	E 233	Thiabendazole
E 202	Potassium sorbate	E 234	Nisin
E 203	Calcium sorbate	E 235	Natamycin
E 210	Benzoic acid	E 239	Hexamethylene tetramine
E 211	Sodium benzoate	E 242	Dimethyl dicarbonate
E 212	Potassium benzoate	E 249	Potassium nitrite
E 213	Calcium benzoate	E 250	Sodium nitrite
E 214	Ethyl *p*-hydroxybenzoate	E 251	Sodium nitrate
E 215	Sodium ethyl *p*-hydroxybenzoate	E 252	Potassium nitrate
E 216	Propyl *p*-hydroxybenzoate	E 260	Acetic acid
E 217	Sodium propyl *p*-hydroxybenzoate	E 261	Potassium acetate
E 218	Methyl *p*-hydroxybenzoate	E 262	Sodium acetate
E 219	Sodium methyl *p*-hydroxybenzoate	E 263	Calcium acetate
E 220	Sulfur dioxide	E 270	Lactic acid
E 221	Sodium sulfite	E 280	Propionic acid
E 222	Sodium bisulfite	E 281	Sodium propionate
E 223	Sodium metabisulfite	E 282	Calcium propionate
E 224	Potassium metabisulfite	E 283	Potassium propionate
E 226	Calcium sulfite	E 284	Boric acid
E 227	Calcium hydrogen sulfite	E 285	Sodium tetraborate
E 228	Potassium hydrogen sulfite	E 290	Carbon dioxide
E 230	Biphenyl (diphenyl)	E 941	Nitrogen
E 231	Orthophenyl phenol	E 1105	Lysozyme
E 232	Sodium orthophenyl phenolate		

4.3 Guidelines to the Granting of Approval

For some time there has been international agreement that food additives, and therefore preservatives for foods, should be permitted for use only if they are technically necessary, entail no risk to health and serve the consumer's interests.

EC requirements are set out precisely in Annex II of the Council Directive dated December 21, 1988 on the harmonization of the laws of the Member States concerning additives authorized for use in foods (89/107/EWG). The requirements of this framework directive, as it is known, also apply to preservatives. Food additives may be permitted for use only:

1) if technical exigency can be demonstrated and the desired aim cannot be achieved with other economically and technically feasible methods;
2) if they entail no risk to the consumer's health at the recommended quantity, insofar as this can be judged from the scientific data available;
3) if the consumer is not misled by their use.

In determining the permissible maximum quantities, account is taken both of the ADI (see Sect. 3.10) and the requirement to permit the preservative in question only in the quantity needed to achieve the desired effect. All preservatives are subject to constant monitoring and may be reclassified in the light of altered conditions of use and new scientific information (see Sect. 3.11).

Preservatives for use in foods have to meet extremely high purity requirements (see Sect. 2.3).

4.4 General Literature

Fortegnelse over godkendte tilsaetningsstoffer til levndesmidler, 1989. Stougaard Jensen, Copenhagen

Dehove RA (1984) Réglementation des produits et services. Qualité, Consommation et Répression des fraudes. Commerce Editions, Paris

Zusatzstoff-Zulassungs-Verordnung of December 22, 1981 (in the current version)

Council Directive dated December 21, 1988 on the harmonization of the laws of the Member States concerning additives authorized for use in foodstuffs (89/107/EWG)

4.5 Specialized Literature

Strahlmann B (1970) Bestrebungen um eine internationale lebensmittelrechtliche Regelung der Lebensmittelzusätze seit Mitte des letzten Jahrhunderts. Lebensm Wiss Technol 3, 1–5

Antimicrobial Action of Preservatives

5.1 General Mechanisms of Action

Food preservatives inhibit not only metabolism but also the growth of bacteria, molds and yeasts.

5.1.1 Inhibitory and Destructive Actions

In practice, a distinction is frequently drawn between a fungistatic or bacteriostatic action (i.e. one that inhibits fungi or bacteria) and a fungicidal or bactericidal action (i.e. one that destroys fungi or bacteria). Close study reveals that such a distinction is unjustifiable. The former two differ from the latter in the death rate of the microorganisms. In the long term the effect of added preservative in food is either to kill the microorganisms or to allow them to grow despite its presence. The governing factor here is the dosage of preservative (see Fig. 3) (Schelhorn 1953).

Depending on the type of preservative used, all the microorganisms are killed within days or weeks at the usual applied concentrations. This is where the crucial difference exists between preservatives and disinfectants. The latter can be employed only if the microorganisms succumb within a very short period of time. The time scale for the killing of microorganisms under the influence of preservatives corresponds to the relationship for a monomolecular reaction.

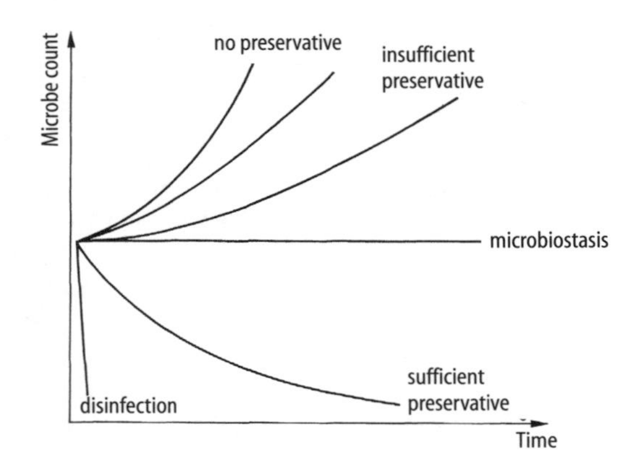

Fig. 3. Trend in microbe count relative to the presence and concentration of antimicrobial substances

$$K = \frac{1}{t} \cdot log \frac{Z_o}{Z_t} \text{ oder } z_t = z_o \cdot e^{-Kt}$$

where
K = death rate constant,
t = time,
z_o = number of living cells at the time the preservative begins to act,
z_t = number of living cells after the time t,

Strictly speaking, this rule is valid only for relatively high dosages of preservatives and a monoclonal population. It also presupposes a closed system, i.e. there must be no attenuation of the preservative, e.g. by evaporation; neither must there be any change in the pH value, nor ingress of further microorganisms, e.g. by reinfection. Even if these requirements are not met completely in the actual practice of food preservation, the foregoing "death rate kinetics" can still be regarded as a good basis for studying the action of preservatives in foods.

Preservatives perform a useful function in foods only if employed in adequate concentrations. The aim is to inhibit the microorganisms in the initial lag-phase and not in the exponential log-phase, since in the latter the dosages of preservative necessary in practice would be far too high. Preservatives are not designed to kill microorganisms in substrates already supporting a massive germ population, i.e. to return putrefying food to an apparently fresh state. Indeed, with the applied concentrations of most preservatives used, this is not even possible. Therefore, preservatives cannot be used to compensate for poor factory hygiene.

5.1.2 Action on Microorganisms

The action of preservatives on the cells of spoilage microorganisms is based on a multiplicity of individual influences. These include not only physical and physicochemical mechanisms but also biochemical reactions. Frequently, several individual factors may produce a cumulative effect, but sometimes only one single stage in a reaction in the microorganism cell is blocked. In the case of spore-forming bacteria, different preservatives develop their inhibitory action at varying stages of spore germination (see Fig. 4) (Gould 1964).

In essence, the antimicrobial action can be explained by the following phenomena (see Fig. 5):

1) influence on the DNA,
2) influence on protein synthesis,
3) influence on enzyme activity,
4) influence on the cell membrane,
5) influence on the cell wall,
6) influence on transport mechanisms for nutrients.

Substances which reduce the water activity of a substrate and thereby impede or inhibit microorganism growth have other criteria of action. Packaging, coatings, oils and some protective gases prevent oxygen from gaining access to foods and

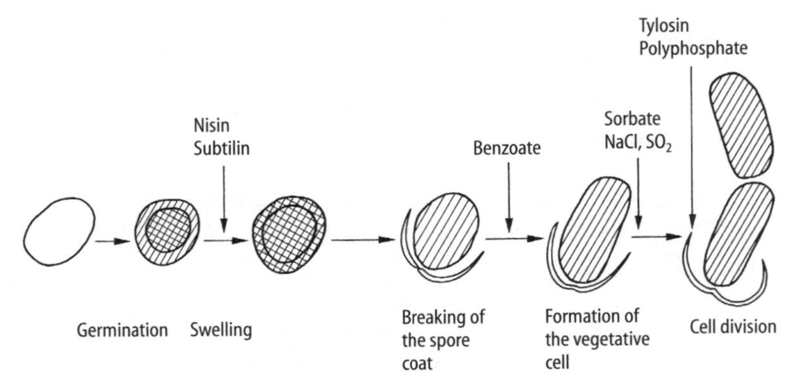

Fig. 4. Possible points of attack for antimicrobial substances during germination of bacterial spores (Gould 1964)

thereby inhibit the growth of aerobic microorganisms. As a rule, none of these materials or substances kills microorganisms; they merely inhibit microorganism growth. If the inhibitory action of the substances diminishes, growth of the contaminants resumes and the product spoils.

One very important aspect is the inhibitory effect of preservatives on enzyme reactions or enzyme and protein synthesis in the microorganism cell. This used to be regarded as the most important factor in explaining the antimicrobial action. Recently it has been increasingly assumed that preservatives act on the cell wall and the cell membrane, their structure and the transport mechanism of nutrients, such as amino acids, to the cell (Freese and Levin 1978, Eklund 1980, Eklund 1981, Gould et al. 1983, Eklund 1985). Lipophilic substances, like most food preservatives, attack the cell membrane, destroying or perforating it (Freese et al. 1973). This increases the flow of protons to the cell. The cell then has to use more energy to compensate for the preservative acid penetrating the neutral interior of the cell (Gould et al, 1983, Eklund 1985) and the potential differences occurring (Eklund 1985).

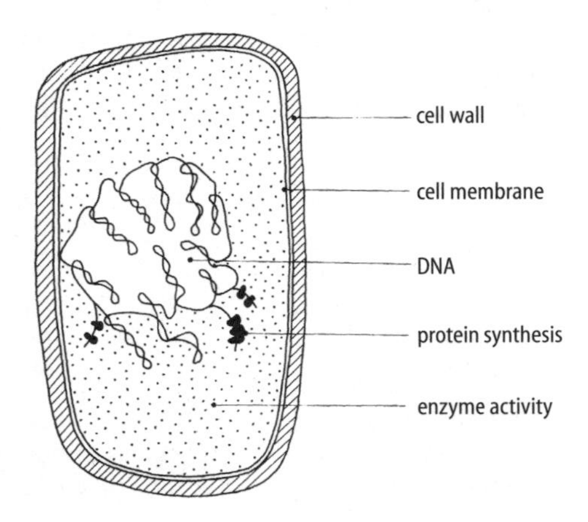

Fig. 5. Possible points of attack for antimicrobial substances in the microorganism cell (Eklund 1981)

Antimicrobial action requires a certain degree of both water solubility and lipid solubility. Microorganism growth occurs solely in the aqueous phase and the preservative must be able to disperse in the water. It also needs to be able to penetrate the hydrophobic cell wall (Branen et al. 1980, Robach 1980).

The efficiency of preservatives is governed by a strict dose-effect relationship. In practical food preservation it is therefore important to employ the preservative in adequate concentrations. The modes of action of the individual preservatives are described in the special chapters on the relevant substances.

5.2 Antimicrobial Spectrum of Preservatives

No preservative has equally powerful effects on molds, yeasts and bacteria. In other words, none of them has a complete spectrum of action against all the spoilage microorganisms likely to occur in food. With most of the preservatives in general use the predominant action is that against yeasts and molds. However, a number of preservatives are bound to be largely ineffectual against certain bacteria because they have virtually no action at the pH optimum of those bacteria, which is frequently in the neutral region. On the other hand, those bacteria cease to grow in any case in the pH range favorable to the relevant preservative; so tests on nutrient media which are optimal for the preservative are certain to be inconclusive.

The literature contains a wealth of data on the inhibitory concentrations of preservatives for particular microorganisms. Some of the data vary substantially. The only values that can be used in comparisons are those obtained under optimum and standardized substrate conditions. Table 7 is intended to provide a survey. Special data concerning the inhibitory concentrations will be found in sections dealing with the individual preservatives.

Table 7. Mode of action of some preservatives on microorganisms

	Bacteria	Yeasts	Molds
Nitrite	++	–	–
Sulfite	++	++	+
Formic acid	+	++	++
Propionic acid	+	++	++
Sorbic acid	++	+++	+++
Benzoic acid	++	+++	+++
p-Hydroxybenzoic acid esters	++	+++	+++
Biphenyl	–	++	++

Key: – ineffective
\+ slightly effective
++ moderately effective
+++ highly effective

5.3 Acquired Resistance to Preservatives

It is known from medical practice that certain microorganisms may develop resistance to active substances in the course of time. Acquired resistance connotes resistance developed by microorganisms when there are repeated passages in the presence of sublethal concentrations, i.e. the ability to tolerate high concentrations of the antimicrobial substance. A process of natural selection takes place – the organisms with little or no resistance are destroyed at low concentrations and those with greater resistance multiply, passing on this trait so that resistance of the population as a whole increases from generation to generation due to the evolutionary pressure exerted by the presence of antimicrobial substances. Acquired resistance occurs primarily among bacteria and usually to antibiotics. As a result, research especially in the medical and crop protection sectors is following entirely new approaches in the search for substances which can reverse resistance (resistance modifiers) (Hollomon 1993).

The total or virtual absence of any inhibitory action by a substance on the capacity of the microorganisms to decompose the active compound is not considered here as resistance in the true sense of the word.

In the long term, acquired resistance of microorganisms to food preservatives would bring about economic damage to the food trade and also a possibility of indirect damage to the health. Consequently, the most important food preservatives have also been carefully studied in experiments with both bacteria and fungi to determine whether resistance is induced (Russell 1991). Acquired resistance has been observed in practical applications of thiabendazole to preserve citrus fruits (Eckert et al. 1981). A documented example of individual strains of microorganisms acquiring resistance to certain preservatives is that of *Listeria monocytogenes* Scott A to nisin. In this particular case fundamental structural and functional membrane changes together with decomposition of the nisin appear to have been the cause (Ming and Daeschel 1993).

In the case of antibiotics not used as food preservatives (penicillin, tetracycline, oxytetracycline and chlorotetracycline), maximum resistance increases by factors between 15 and 150, and in the case of streptomycin by a factor as high as 1670 (Lukas 1964). The resistance of *Escherichia coli* to antibiotics can actually be largely eliminated by the addition of sorbic acid, even in sublethal doses (Athar and Winner 1971). The possibility of acquired resistance is the main reason why antibiotics are not used in food preservation. Should there, however, be exceptional cases in which antibiotics are employed for certain purposes in food preservation, only those not being concurrently used for therapeutic purposes will be accepted.

5.4 Combinations of Preservatives with One Another

In medicine the practice of combining various substances with one another is sometimes employed in order to enhance or modify their action compared with that of the individual constituents. Following this practice, combinations of preservatives have been evolved simply by trial and error. The normal reason for using such combinations in food technology is to provide the following advantages (Rushing and Senn 1963):

1) broader spectrum of action,
2) increased antimicrobial action,
3) lower concentration of individual substances.

5.4.1 Broader Spectrum of Action

No preservative is active against all the spoilage microorganisms present in a foodstuff. An attempt is therefore made to compensate for this deficiency by combining various preservatives having different spectra of action. Theoretically it is possible for a combination to have a spectrum of action different from that of the two individual constituents; and if this happens the spectrum of action may include microorganisms not inhibited by the individual constituents or else inhibited by them only in extreme concentrations (Rehm 1959 b). Thus, for instance, combinations of sorbic acid and benzoic acid inhibit a number of bacteria strains better than either sorbic acid or benzoic acid alone (Rehm 1959 a, Rehm and Stahl 1960).

For practical purposes, interest focuses primarily on the combination of benzoic acid and sorbic acid, both effective mainly against yeasts and molds, with antibacterial substances. The most important product used as the antibacterial constituent in combination with either benzoic acid or sorbic acid is sulfur dioxide, especially when employed in conjunction with foods of plant origin, where its enzyme-inhibiting and antioxidative properties are exploited simultaneously.

5.4.2 Changes in Antimicrobial Action

There are three possible means whereby the action may be changed by combining two or more preservatives:

1) addition or additive effect,
2) synergism or synergistic effect,
3) antagonism or antagonistic effect.

The term additive effect denotes that the effects of the individual substances are simply added together. Synergistic effect is the expression used when the inhibitory action of the combination is reached at a concentration lower than that of the constituent substances separately. An antagonistic effect is the opposite of this latter, i.e. one where the mixture concentration required is higher than that of the individual constituents.

Combining two or more synergistically acting preservatives with one another enables the total content of preservative in the food to be reduced or makes for a reduction in any organoleptic side effects (Rehm 1959 b). No generally valid statements can be made about the actions of any particular preservative on other preservatives (Rehm 1959 a, Rehm and Stahl 1960, Rehm 1960, Rehm et al. 1964 a, Rehm et al. 1964b). Any synergistic effects observed in laboratory tests are so weak as to have no significance for practical food preservation. It was mainly for commercial reasons that most combination products formerly marketed in large numbers were preferred to straight preservatives.

What may be regarded as another means of intensifying the action is to combine a conventional preservative having a long-term action, such as sorbic acid, with a substance whose action is rapid but of less duration, such as diethyl pyrocarbonate (Harrington and Hills 1966). The latter-named preservative ensures that the germs present are killed rapidly, while the sorbic acid affords protection against reinfection.

In principle, a beneficial effect is obtained by using preservatives in the narrow sense of the term together with substances that counter dissociation, such as acids, or ones that produce osmotic effects and reduce the water activity, e.g. common salt or sugar.

5.5 Combination of Preservatives with Physical Measures

In just the same way as several preservatives can be employed together, so it is frequently useful to combine preservatives with physical methods of food preservation, such as heating, refrigeration, irradiation, drying or modern processes, such as high-pressure treatment, ohmic heating or pulse electric field techniques. This enables undesired side-effects of the individual methods to be avoided or reduced. Another important fact is that the use of preservatives is far less expensive in terms of energy than conventional physical methods of food preservation.

The combined use of several preservation methods, possibly physical and chemical, and of different preservatives is an age-old practice. As long ago as the days of the ancient Egyptians there was the practice of mummification. Though not a method of food preservation, it combined lowering the water activity, raising the pH value to a range within which bacterial growth was severely restricted, and the use of antibacterial substances (Chirife et al. 1991). Ultimately, acidification also involves a combination of lowering the pH value by adding or forming acid with the use of common salt whose action is to lower the water activity. In a wider sense, smoking is another example of the combined use of preservatives and physical measures, since the antimicrobial action of smoke constituents is combined with the smoke's drying effect. For decades, the preservative effect of substances has been improved by cold storage of the foods to which they are added, as in the case of benzoic acid in fish preserves.

In recent years the effects of combining preservatives with physical measures have been investigated more systematically by Leistner, who developed the concept of "hurdle technology" to describe these effects (Leistner 1978, Leistner et al. 1981, Leistner 1992, Leistner 1995). This states that several hurdles (inhibitory factors), even if individually unable to inhibit microorganisms, will nevertheless together prevent microorganism growth if incorporated into a substrate in sufficient number and height. Figure 6 illustrates the principles of the hurdle concept in simplified form. The most important hurdles (inhibitory factors) for microorganisms are:

1) Initial microbe count: Should be as low as possible.
2) Storage temperature: Should be as low as possible.
3) pH: Some bacteria are inhibited by a low pH value alone. The lower the pH value, the more effective many preservatives are.
4) Water activity: Should be as low as possible if capable of variation.
5) Oxygen ingress: Should be as low as possible.
6) Degree of heating: Should be as high as possible.
7) Preservatives: Should be present in adequate concentration.

5.5.1 Combination of Preservatives with the Use of Heat

In the presence of preservatives the temperature/time values required to kill the microorganisms are lower than in the absence of preservatives. In other words, the microorganisms are killed more swiftly in the presence of preservatives than at the

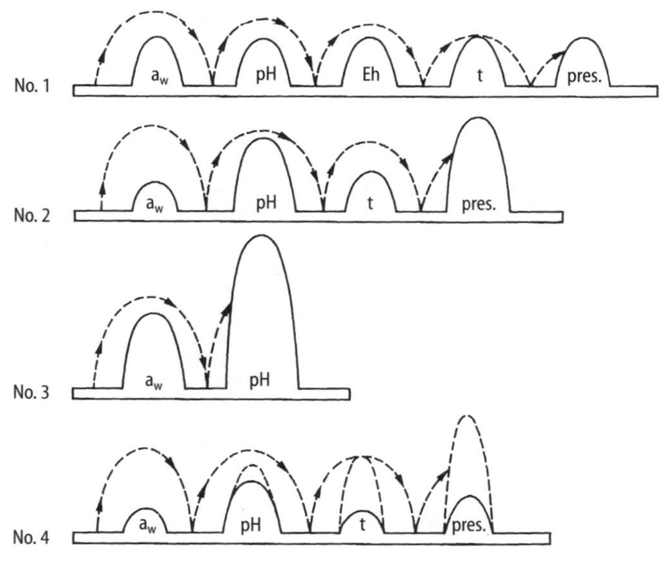

Fig. 6. Hurdle technology exemplified by four different food products (Leistner et al. 1981). In No. 1 and No. 2 only the preservative hurdle is sufficiently high. In No. 3 the pH is the decisive hurdle. In No. 4 none of the hurdles is effective.

a_w water activity
pH pH value
Eh Redox potential
t effect of temperature (heating, cooling)
pres. presence of preservatives

same temperature in their absence. These relationships have been confirmed in laboratory tests with many strains of bacteria and most of the usual preservatives, e.g. with yeasts and benzoic acid or salicylic acid (Ko Swan Djien et al. 1972, Beuchat 1981 a, 1982, 1983), yeasts and pimaricin (York 1966), yeasts and sorbic acid (Ko Swan Djien 1973, Beuchat 1981 a, 1982, 1983, Sankaran and Leela 1983), molds and benzoic acid (Beuchat 1981 b), molds and sorbic acid (Beuchat 1981 b) as well as salmonella and sorbic acid (Lerche and Linke 1958). Reducing the water activity, by adding common salt or sucrose for example, increases the resistance of yeast cells to the effect of heat (Doyle and Marth 1975, Beuchat 1981 c). The addition of sorbic acid or benzoic acid, however, restores the sensitivity of the yeasts even if the water activity is reduced (Beuchat 1981 c, 1981 d).

The synergy which doubtless exists between the use of heat and preservatives is not sufficient to be of any great practical importance.

5.5.2 Combination of Preservatives with Refrigeration Treatment

In principle, the combination of preservatives with refrigeration treatment follows the same rules as their combination with heat. If cold storage is employed, it is sufficient to use concentrations of preservatives which are in fact inadequate to prevent spoilage at room temperature (Schelhorn 1953, Schelhorn 1954).

5.5.3 Combination of Preservatives with Irradiation

The use of preservatives enables the radiation dosage to be reduced when foodstuffs are exposed to ionizing rays. In vitro experiments and practical tests have consistently revealed synergistic effects between the ionizing rays and preservatives, e.g. in fish products (Maha et al. 1980). Preservatives such as sorbic acid also minimize changes in foods caused by irradiation, for example distorted flavors or discoloration, and thereby function as radioprotectors (Thakur and Arya 1993). Preservatives may be combined with ionizing radiation only if they have adequate resistance to ionizing rays (Thakur et al. 1990). The antimicrobial action of ultraviolet rays is also enhanced by the presence of preservatives, e.g. by sorbic acid (Harrington and Hills 1968).

5.6 Preservation Against Toxin-Forming Microorganisms

Whereas, in former times, preservatives were used in foods almost exclusively for economic reasons, they are now being employed increasingly as a protection against toxin-forming microorganisms, bacteria and molds. The use of preservatives is a prophylactic measure since, by preventing spoilage of food, one avoids the risk of toxin formation from the outset. Such prophylaxy is of added importance inasmuch as toxins which have already formed cannot generally be removed from foodstuffs or feed afterwards.

Preservatives have long been used to protect against toxin-forming bacteria, though this effect was initially largely unintentional. Nitrite and nitrate not only redden meat and impart a cured flavor; the main action of nitrite is actually to inhibit the development of clostridia and thus the formation of botulism. Because of disagreements over the toxicological characteristics of nitrite, extensive studies were conducted in the early nineteen-eighties on the action of other preservatives used in combination with smaller nitrite quantities or without nitrite against toxin-forming bacteria, especially clostridia. During those studies, very good effects were achieved with sorbic acid, but it has failed to gain general acceptance despite these test results because, unlike nitrite, it does not redden meat or impart a cured flavor.

Preservatives are also used as protection against toxin-forming fungi to reduce the risk posed by mycotoxins (Frank 1974, Lück 1981). Hard cheese, dry sausage, fruit products, and baked goods are the foods exposed to the greatest risk and also those which, for technical reasons, are the most easily protected by preservatives against the risk associated with mycotoxins. Roughly one-third of the molds that grow spontaneously on American Emmentaler have proved to be mycotoxin-forming fungi (Bullerman 1976).

In particular, the action of sorbic acid on mycotoxin-forming fungi has been thoroughly investigated (Groll and Lück 1970, Wallhäußer and Lück 1970, Reiß 1976, Chipley et al. 1981, Ray 1982, Bullerman 1983, Bullerman 1984a, Bullerman 1984b, Tong and Draughon 1985). Mycotoxin formation is, however, also suppressed by other preservatives, such as propionic acid, benzoic acid and natamycin (Ray and Bullerman 1982, Bullerman et al. 1984, Liewen and Marth 1985). Sorbic acid has a more powerful effect against mycotoxin formation than against the mold growth as such (Groll and Lück 1970, Bullerman 1983, Bullerman 1984a). It also inhibits aflatoxin synthesis in resting cultures of *Aspergillus parasiticus*, though not in growing cultures (Yousef and Marth 1983). The same is true of propionates (Chourasia 1993). If the mold is present in high concentrations, it grows on suitable culture solutions even in the presence of 3000 ppm potassium sorbate. Even then, however, potassium sorbate inhibits the formation of mycotoxins (Liewen and Marth 1985). Subthreshold dosages of propionic acid (Bauer et al. 1983) and sorbic acid (Bauer et al. 1983, Gareis et al. 1984) may stimulate aflatoxin formation.

Sulfites destroy aflatoxins formed under certain conditions (Tabata et al. 1994) and could theoretically be used to "decontaminate" infected products. However, such a procedure is inadvisable. It is preferable to add preservatives, thereby preventing mold growth and consequent aflatoxin formation from the outset.

5.7 Influences of Substrate Factors on the Action of Preservatives

The action of preservatives is influenced by physico-chemical aspects of the substrate and by certain constituents of the foods to be preserved, e.g. by substances that change the pH value or water activity or ones that selectively absorb the preservative. Other influences on the action of preservatives are those exerted by natural food ingredients which have their own antimicrobial action. Some of these substrate factors increase the action of preservatives, whereas others reduce it. For these reasons the effective applied concentration of a preservative in a food is rarely the same as the minimum inhibitory concentration determined in vitro.

5.7.1 Influence of the pH Value

In the case of preservatives subject to dissociation in aqueous systems on account of their structure, either the hydrogen ions released in solution or simply the undissociated acid constituent may be responsible for the antimicrobial action.

One typical preservative that acts by way of the released hydrogen ions is acetic acid. A predominant feature in this instance is the action of the acid as such, i.e. the reduction in the pH value, which makes it impossible for some microorganisms, especially bacteria, to live. Preservatives of this type have to be employed in relatively high concentrations, since quantities of acid amounting to one or more percent are necessary to achieve any appreciable reduction in the pH of buffered substrates, such as the majority of foods.

A typical representative of the second group is sorbic acid. In this case, it is in practice nearly always the undissociated acid constituent that has an antimicrobial action. It is not the hydrophilic acid anions but only the undissociated compounds, because of their lipophilic characteristics, that can migrate through the semipermeable cell membrane of the microorganisms and exert their action – usually of an enzyme-inhibiting form – in the interior of the cell. Unlike the first-named group, preservatives of this type act in concentrations of far less than one percent.

The content of undissociated acid declines with an increase in the pH value. Preservatives susceptible to dissociation are thus more effective the lower the pH value of the substrate. The proportion of undissociated constituent can be calculated from the dissociation constant by the following formula (Rahn and Conn 1944):

$$a = \frac{[H^{\cdot}]}{[H^{\cdot}] + D}$$

a = amount of undissociated acid
$[H^{\cdot}]$ = hydrogen ion concentration
D = dissociation constant

Table 8 gives the amount of undissociated acid for the most important preservatives based on carboxylic acid in the pH range from 3 to 7. It can be seen from the table that for all practical purposes these preservatives cease to have any action whatever or retain only a very slight action in the neutral pH range. In this range only the non-dissociating compounds such as boric acid or *p*-hydroxybenzoic acid esters can be used. Differences important for practical conditions are evidenced in the weakly acid region of pH 5 to 6, into which category many foodstuffs fall. Sorbic acid and propionic acid are still suitable in this range, whereas benzoic acid and formic acid are in most cases ineffectual.

Theoretically it would be possible to compensate for the small proportion of undissociated acids by increasing the total addition of the preservative in question. However, in some instances this approach would encounter food law difficulties

Table 8. Undissociated proportions of preservative acids at various pH values

Preservative	Dissociation constant	pK value	Percentage of undissociated acid at pH								
			3.0	3.5	4.0	4.5	5.0	5.5	6.0	6.5	7.0
Sulfur dioxide	$1.54 \cdot 10^{-2}$	1.81	6	2	0.6	0.2	0.06	0.02	0.01	0	0
Salicylic acid	$1.07 \cdot 10^{-3}$	2.97	48	23	9	3	1	0.3	0.1	0.03	0.01
Formic acid	$1.77 \cdot 10^{-4}$	3.75	85	64	36	15	5	1.8	0.6	0.2	0.06
p-Chlorobenzoic acid	$9.3 \cdot 10^{-5}$	4.03	92	77	52	25	10	3.3	1.1	0.3	0.1
Benzoic acid	$6.46 \cdot 10^{-5}$	4.18	94	83	61	33	13	5	1.5	0.5	0.15
p-Hydroxybenzoic acid	$3.3 \cdot 10^{-5}$	4.48	97	91	75	49	23	9	2.9	1.0	0.3
Acetic acid	$1.76 \cdot 10^{-5}$	4.75	98	95	85	64	36	15	5.4	1.8	0.6
Sorbic acid	$1.73 \cdot 10^{-5}$	4.76	98	95	85	65	37	15	5.5	1.8	0.6
Propionic acid	$1.32 \cdot 10^{-5}$	4.88	99	96	88	71	43	19	7.0	2.3	0.8
Dehydroacetic acid	$5.30 \cdot 10^{-6}$	5.27	100	98	95	86	65	37	15.9	5.6	1.9
Hydrogen sulfite	$1.02 \cdot 10^{-7}$	6.99	100	100	100	100	99	97	91	76	50
Boric acid	$7.3 \cdot 10^{-10}$	9.14	100	100	100	100	100	100	100	100	99

and would in any case be impracticable since the additions necessary would be excessively large and sometimes far in excess of the taste threshold value.

Sulfurous acid is in a category of its own, since it can exist as dissolved SO_2 gas, as undissociated sulfurous acid H_2SO_3, as a hydrogen sulfite ion HSO_3^- and as a sulfite ion SO_3^{2-}. The antimicrobial action of the individual forms differs widely, yet sulfurous acid still displays a certain antimicrobial action, even in the medium pH range, owing to the presence of hydrogen sulfite ions (Rehm and Wittmann 1962).

For the reasons stated, a shift in the pH value of a foodstuff to the acid range will, in principle, improve the action of the preservatives. Indeed, such a measure is also of practical value inasmuch as a highly acid environment prevents the growth of a number of bacteria strains. Nevertheless, the changes in the pH value of a food are subject to certain limits for reasons of taste.

There is evidence to suggest that preservative acids have an antimicrobial action not only in the form of undissociated molecules but in the form of dissociated molecules also, though the action of the latter is much lower than that of the undissociated acids. In the case of sorbic acid the difference is a factor of about 100 (Rehm and Lukas 1963, Eklund 1983).

5.7.2 Influence of the Partition Coefficient

The partition coefficient, defined by the ratio of solubility values in the fatty and aqueous phase, is of practical importance for the preservation of foods with a high fat content, such as emulsions. In systems of this type, microbial growth occurs exclusively in the aqueous phase and consequently the proportions of preservatives that migrate into the fatty phase are of no importance. Hence, the most highly rated preservatives are those which display the lowest partition coefficients. Owing to the differing solubilities of the preservatives in the various fats, the partition coefficient depends on the nature of the fat concerned.

Common salt, sugar and other dissolved substances increase the partition coefficient because they reduce the solubility in the aqueous phase ("salting-out effect"). The partition coefficient declines with a rise in the pH value because only the undissociated constituents of the preservatives dissolve in the fatty phase. Table 9 provides a survey of the partition coefficients for various preservatives between

Table 9. Partition coefficient of preservatives

	Edible oil	Octanol
Propionic acid	0.17	
Sorbic acid	3.0	21
p-Hydroxybenzoic acid methyl ester	5.8	91
Benzoic acid	6.1	51
p-Hydroxybenzoic acid ethyl ester	26.0	
p-Hydroxybenzoic acid propyl ester	87.5	1010

edible oils (Lubieniecki-von Schelhorn 1967a and b) or octanol (Schmidt and Franck 1993) and water.

5.7.3
Influence of the Water Activity

The addition of substances that reduce the water activity of a foodstuff (the a_w value) has a fundamentally beneficial effect on the action of preservatives. The most important substances in this respect are common salt, sugar, glycerin and glycols. Details concerning the limiting values of the water activity for some important microorganisms occurring in foodstuffs are provided in Table 10 (Lück 1973, Leistner et al. 1981). This table shows essentially that most bacteria are capable of growing only in conjunction with high water activities. There are, however, many molds and yeasts still quite capable of development at a_w values below 0.85.

The toxin formation of microorganisms is also linked to the water activity, although the a_w value is not the only limiting factor. Mycotoxin formation is possible even at a_w values of around 0.76; the formation of penicilloic acid by *Aspergillus ochraceus* is one such example (Leistner et al 1981, Beuchat 1983). *Clostridium botulinum* type B is the most sensitive strain and does not form any toxin at a_w values below 0.94 (Leistner et al. 1981). The formation of Enterotoxin A by *Staphylococcus aureus* ceases at a_w values below 0.87 (Leistner et al. 1981).

5.7.4
Influence of Other Physico-Chemical Substrate Factors

Microorganisms impose a variety of demands on other substrate properties such as redox potential and partial oxygen pressure (Lubieniecki-von Schelhorn 1975). Consequently, all additions and ingredients of the foods that change these factors have an influence on the action of added preservatives. A preservative typical for its action on the redox potential is sulfur dioxide. The partial oxygen pressure of a system is influenced, amongst other factors, by the preservatives common salt, carbon dioxide and nitrogen (Lubieniecki-von Schelhorn 1975).

5.7.5
Influence of Food Ingredients

From among the ingredients of foodstuffs themselves or the additions required for reasons other than preservation, the substances that most influence the action of preservatives are common salt, carbohydrates and alcohol.

The common salt content of a number of foodstuffs intensifies the action of preservatives chiefly through the removal of water by osmosis and the ensuing reduction in the water activity. Common salt also induces swelling and thereby renders many microorganisms more susceptible to attack by preservatives. Finally, it supports the action of preservatives by a direct influence on enzymes. In emulsions, however, common salt has an unfavorable influence on preservation because it increases the partition coefficient of preservatives.

Table 10. Minimum levels of water activity for various important microorganisms occurring in foods (Leistner et al. 1981)

Bacteria		Yeasts	Molds
0.98	*Clostridium* (1), *Pseudomonas*[a]	–	–
0.97	*Clostridium* (2), *Pseudomonas*[a]	–	–
0.96	*Flavobacterium, Klebsiella, Lactobacillus*[a], *Proteus*[a], *Pseudomonas*[a], *Shigella*	–	–
0.95	*Alcaligenes, Bacillus, Citrobacter, Clostridium* (3) *Enterobacter, Escherichia, Propionibacterium, Proteus, Pseudomonas, Salmonella, Serratia, Vibrio*	–	–
0.94	*Bacillus*[a], *Clostridium* (4), *Lactobacillus, Microbacterium, Pediococcus, Streptococcus*[a], *Vibrio*	–	*Stachybotrys*
0.93	*Bacillus* (5), *Micrococcus*[a], *Lactobacillus*[a], *Streptococcus*	–	*Botrytis, Mucor, Rhizopus*
0.92	–	*Pichia, Rhodotorula, Saccharomyces*[a]	–
0.91	*Corynebacterium, Streptococcus*	–	–
0.90	*Bacillus* (6), *Lactobacillus*[a] *Micrococcus, Pediococcus, Staphylococcus* (7), *Vibrio*[a]	*Hansenula, Saccharomyces*	–
0.88	–	*Candida, Debaryomyces, Hanseniaspora, Torupolis*	*Cladosporium*
0.87	–	*Debaryomyces*[a]	–
0.86	*Micrococcus*[a], *Staphylococcus* (8), *Vibrio* (9)	–	–
0.84	–	–	*Alternaria, Aspergillus*[a], *Paecilomyces*
0.83	*Staphylococcus*[a]	*Debaryomyces*[a]	*Penicillium*[a]
0.81		*Saccharomyces*[a]	*Penicillium*[a]
0.79	–	–	*Penicillium*[a]
0.78			*Aspergillus*[a], *Emericella*
0.75	*Halobacterium, Halococcus*	–	*Aspergillus*[a], *Wallemia*
0.70	–	–	*Aspergillus*[a] *Chrysosporium*
0.62	–	*Saccharomyces*[a]	*Eurotium*[a]
0.61	–	–	*Monascus (Xeromyces)*

[a] Various strains;
(1) *Clostridium botulinum* type C;
(2) *Clostridium botulinum* type E and various strains of *C. perfringens*;
(3) *Clostridium botulinum* type A and B as well as *C. perfringens*;
(4) various strains of *C. botulinum* type B;
(5) various strains of *Bacillus stearothermophilus*;
(6) *Bacillus subtilis* under certain conditions;
(7) *Staphylococcus aureus* in anaerobic growth conditions;
(8) *Staphylococcus aureus* in aerobic growth conditions;
(9) various strains of *Vibrio costicolus*.

From among the carbohydrates, sugars have the greatest importance as substrate factors influencing the action of preservatives. The presence of sugars as nutrients readily convertible in low concentrations and thus promoting the growth of microorganisms is the sole reason why preservatives need be employed at all in a number of foods. In higher concentrations, however, sugars inhibit microorganism growth by reducing the water activity in a manner similar to that of common salt. Sugars also closely resemble common salt in the way they influence the partition coefficient, although their action is weaker. In principle, alcohol enhances the action of preservatives.

Chemically reactive preservatives may become linked to ingredients of foodstuffs and thereby partially or completely lose their efficacy in certain circumstances. If reactions of this type are likely, the preservative must generally be employed in greater dosages in order to compensate. This is especially important in the case of sulfur dioxide, which reacts with aldehydes and glucose (Rehm et al. 1964). In wine this reaction is desirable since it results in the chemical linkage of acetaldehyde, a fermentation byproduct which affects the taste. Nitrite may also react with food ingredients. Thus under certain conditions nitrite and amines may form carcinogenic *N*-nitroso compounds, e.g. nitrosamines. Further particulars on the possible reactions of preservatives with food ingredients are discussed in the chapters on the individual preservatives themselves.

5.8 Decomposition of Preservatives

Usually, food preservatives are chemically stable substances and thus unlikely to decompose within the storage times customary for the foodstuffs in question. Exceptions among inorganic preservatives are nitrites, sulfites, hydrogen peroxide and ozone. Among the organic compounds the exceptions are dicarbonate esters and antibiotics (Sharp 1976).

Among some of these substances, decomposition is unavoidable and indeed constitutes part of their effect, as in the case of hydrogen peroxide, which acts by releasing oxygen. In other cases decomposition is desired because it ensures that the final product contains no preservatives, e.g. dicarbonate esters.

Besides chemical decomposition, a number of preservatives can be decomposed by microorganisms. This is true in particular of organic compounds, which can be utilized by a number of microorganisms as a source of carbon. Methylparaben can be decomposed by *Pseudomonas aeruginosa* (Parrof 1968), sorbic acid by Penicillium strains (Kinderlerer and Hatton 1990) and other molds (Lück 1972). Decomposition may occur if the relevant preservative is ineffectual against the microbe in question. It is also possible if the preservative concentration is inadequate for the microbe content, as in the case of heavily contaminated foods or where there is incipient microbial spoilage. In principle, therefore, when spoilage has once set in, it is impossible to arrest it with preservatives and return the food to an apparently fresh condition. Hence, the consumer of foods kept fresh with preservatives capable of microbiological decomposition has the assurance that the starting materials processed were microbiologically in perfect condition from the outset.

5.9
Test Methods for Preservatives

To test the efficacy of preservatives for foods, it is possible to use two methods which are normally employed in succession, i.e. the nutrient medium test and the practical trial.

5.9.1
Nutrient Medium Test

In the case of the nutrient medium test, which comes into its own primarily when unknown active substances are involved, a nutrient medium is treated with graduated quantities of the substance to be tested and then inoculated with a defined population of a pure culture of the microorganism to be tested. After a few days' incubation it is possible to ascertain, from the varying development of the cultures, the active substance concentration just sufficient to inhibit growth. The influences of substrates are important in this connection. It is also important to have optimum growth conditions for the microorganism in question and also optimum conditions of action for the preservative tested. Evaluation is carried out with liquid nutrient medium by measuring the turbidity in a spectrophotometer or, in the case of fungi, determining the increase in mycelial weight.

The agar cup test, which is familiar from bacteriology, is only of limited importance in the testing of food preservatives as it is based on the assumption that the active substances are required to inhibit large numbers of microbes. In food preservation, however, the numbers present should not be large. Hence, the aim is to add preservatives to control spoilage microbes present in small numbers. The tests for cosmetics and pharmaceuticals described in pharmacopeia and the CTFA guidelines (Leimbeck 1991) are also of very limited importance for food products. The required storage stability and normal infection and post-infection conditions associated with cosmetics are entirely different from those in the food sector.

Although the nutrient medium test conducted in vitro is highly accurate from a microbiological viewpoint, it is of interest only as a guide in practical use since food can influence preservatives in a number of ways which this test is unable to simulate. Moreover, a food is by no means always as "ideally contaminated" as a nutrient medium, since in practice what occur are a large number of different microorganism strains and not one single strain on its own.

5.9.2
Practical Trials

Nutrient medium tests must, in principle, be backed up by practical trials. The food to be tested is employed in these trials as the nutrient medium. The preservative to be tested is added in differing concentrations to various batches of the food intended for preservation. Some of the samples in the standard food packaging are inoculated with microorganism strains known to cause spoilage in the particular type of food involved. Of these samples a number are stored under normal storage conditions, i.e. mainly at normal storage temperature and in some cases under more stringent conditions, chiefly at elevated temperature. At certain intervals of time, which have to be governed by the customary and/or desired storage time for the food, the food is tested for its keeping properties. Determination of its organoleptic properties is insufficient. In addition, the change in chemical and physical characteristic data will be monitored, the choice of data depending on the type of foodstuff in question. A factor of major importance is the change in the microbe count as determined by microbiological methods.

Another difficulty is the task of inoculating with the correct microorganisms. What matters here is not only to inoculate with suitable types of microorganism, i.e. types that act as spoilage microorganisms in actual practice, but also that the conditions under which the work is carried out should resemble those of actual practice as closely as possible, and this is not always simple. Preservatives are designed not to combat unduly high microbe counts but to deal with a relatively small microbe population such as may be expected on food produced under hygienic conditions. Some preservatives can be decomposed by microorganisms if present in unusually large numbers. If tests to determine the action of a preservative are carried out with unduly high microbe counts, they may make the preservative appear ineffectual even if it has a satisfactory antimicrobial action in foods having the usual level of contamination encountered in everyday practice.

5.10 Selection of a Suitable Preservative

It is not possible for every food to be preserved with any preservative that is in principle suitable. When assessing the suitability of a preservative in a particular case the preservative must be examined in the light of the following requirements, which it should, if possible, satisfy.

1) The preservative must be harmless to health.
2) Even in its pure form the preservative should be toxicologically harmless and ecologically compatible during use and processing.
3) The preservative should have the widest possible spectrum of action. Its action against the microorganisms expected or normally occurring in or on foods must be adequate under the conditions prevailing in the food concerned (pH, a_w value and other substrate factors) (see Sect. 5.7).
4) The preservative should be effective against toxin-forming microorganisms and if possible inhibit toxin formation more effectively than the growth of microorganisms.
5) There should be no acquired resistance to the preservative.
6) The preservative should cause little or no disruption of desirable microbiological processes occurring in the food, e.g. yeast fermentation in baked goods, lactic acid fermentation in pickled vegetables and cheese ripening.
7) The preservative should remain stable in the food until the food is consumed.
8) The preservative should not react with food constituents or at least not until its antimicrobial action is no longer required.

9) The preservative should not react with packaging or be absorbed by such materials.
10) The preservative should cause minimum impairment to the sensory characteristics of the food, e.g. odor, taste, color and texture.
11) The preservative should be easy to use. In water-containing foods some degree of water-solubility is required.
12) The preservative should be as inexpensive as possible so as not to produce an unreasonable increase in the price of the food. It is usually more economical to use preservatives, even relatively expensive ones, than to employ physical preservation methods, e.g. thermal processes and irradiation.
13) The preservative must be approved for its intended use under food regulations or its approval must be imminent.
14) The quality and purity of the preservative must comply with national and international standards and requirements.

5.11 Literature

Athar MA, Winner HI (1971) The development of resistance by *Candida* species to polyene antibiotics in vitro. J Med Microbiol 4, 505–517

Bauer J, Montgelas A, von Gedek B (1983) Aflatoxin B_1 production in presence of preservatives antimicrobial agent. Proc Int Symp Mycotoxins, p. 249–255

Beuchat LR (1981a) Synergistic effects of potassium sorbate and sodium benzoate on thermal inactivation of yeasts. J Food Sci 46, 771–777

Beuchat LR (1981b) Influence of potassium sorbate and sodium benzoate on heat inactivation of *Aspergillus flavus*, *Penicillium puberulum* and *Geotrichum candidum*. J Food Protect 44, 450–454

Beuchat LR (1981c) Combined effects of solutes and food preservatives on rates of inactivation of and colony formation by heated spores and vegetative cells of molds. Appl Environm Microbiol 41, 472–477

Beuchat LR (1981d) Effects of potassium sorbate and sodium benzoate on inactivating yeasts heated in broths containing sodium chloride and sucrose. J Food Protect 44, 765–769

Beuchat LR (1982) Thermal inactivation of yeasts in fruit juices supplemented with food preservatives and sucrose. J Food Sci 47, 1679–1682

Beuchat LR (1983) Influence of water activity on growth, metabolic activities and survival of yeasts and molds. J Food Protect 46, 135–141

Branen AL, Davidson PM, Katz B (1980) Antimicrobial properties of phenolic antioxidants and lipids. Food Technol 34:5, 42

Bullerman LB (1976) Examination of Swiss cheese for incidence of mycotoxin producing molds. J Food Sci 41, 26–28

Bullerman LB (1983) Effects of potassium sorbate on growth and aflatoxin production by *Aspergillus parasiticus* and *Aspergillus flavus*. J Food Protect 46, 940–946

Bullerman LB (1984a) Effects of potassium sorbate growth on patulin production by *Penicillium patulum* and *Penicillium roqueforti*. J Food Protect 47, 312–315 and 320

Bullerman LB (1984b) Inhibition of ochratoxin production by sorbate. J Food Protect 47, 820–821

Bullerman LB, Schroeder LL, Park K-Y (1984) Formation and control of mycotoxins in food. J Food Protect 47, 637–646

Chipley JR, Story LD, Todd PT, Kabara JJ (1981) Inhibition of Aspergillus growth and extracellular aflatoxin accumulation by sorbic acid and derivatives of fatty acids. J Food Safety 2, 109–120

Chirife J, Favetto G, Ballesteros S, Kitic D (1991) Mummification in ancient Egypt: an old example of tissue preservation by hurdle technology. Lebensmittel-Wiss u. Technol 24, 9–11

Chourasia M (1993) Growth, sclerotia and aflatoxin production by *Aspergillus parasiticus*: influence of food preservatives. Lett Appl Microbiol 17, 204–207

Doyle MP, Marth EH (1975) Thermal inactivation of conidia from *Aspergillus flavus* and *Aspergillus parasiticus*. II. Effects of pH and buffers, glucose, sucrose and sodium chloride. J Milk Food Technol 38, 750–758

Eckert JW, Bretschneider BF, Ratnayake M (1981) Investigations on new postharvest fungicides for citrus fruits in California. Proc Int Soc Citriculture 2, 804–810

Eklund T (1980) Inhibition of growth and uptake processes in bacteria by some chemical food preservatives. J Appl Bacteriol 48, 423–432

Eklund T (1981) Chemical food preservations – some basic aspects and practical considerations. Applied food science in food preservation (paper presented in London on November 18, 1981)

Eklund T (1983) The antimicrobial effect of dissociated and undissociated sorbic acid at different pH levels. J Appl Bacteriol 54, 383–389

Eklund T (1984) The antimicrobial action of some food preservatives at different pH levels. In: Kiss I, Deák T, Incze K: Microbial Associations and Interactions in Food. Budapest, p. 441–445

Eklund T (1985) The effect of sorbic acid and esters of *p*-hydroxybenzoic acid on the protonmotive force in *Escherichia coli* membrane vesicles. J Gen Microbiol 131, 73–76

Frank HK (1974) Aflatoxine. Bildungsbedingungen, Eigenschaften und Bedeutung für die Lebensmittelwirtschaft. Behr, Hamburg

Freese E, Scheu CW, Galliers E (1973) Function of lipophilic acids as antimicrobial food additives. Nature 241, 321–325

Freese E, Levin BC (1978) Action mechanisms of preservatives and antiseptics. Dev Ind Microbiol 19, 207–227

Gareis M, Bauer J, Montgelas A, von Gedek B (1984) Stimulation of aflatoxin B_1 and T-2 toxin production by sorbic acid. Appl Environm Microbiol 47, 416–418

Gould GW (1964) Effect of food preservatives on the growth of bacteria from spores. In: Microbial Inhibitors in Food. Almqvist & Wiksell, Stockholm, p. 17–24

Gould GW, Brown MH, Fletcher BC (1983) Mechanisms of action of food preservation procedures. Soc Appl Bateriol Symp Series 11, 67–84

Groll D, Lück E (1970) Wirkung von Sorbinsäure und Sorboylpalmitat auf die Aflatoxinbildung im Brot. Z Lebensm Unters Forsch 144, 297–300

Harrington WO, Hills CH (1966) Preservative effect of diethyl pyrocarbonate and its combination with potassium sorbate on apple cider. Food Technol 20, 1360–1362

Harrington WO, Hills CH (1968) Reduction of the microbial population of apple cider by ultraviolet irradiation. Food Technol 22, 1451–1454

Hollomon DW (1993) Pesticide Resistance. Chem Ind (London) 892–893

Kinderlerer J, Hatton P (1990) Fungal metabolites of sorbic acid. Food Add Contam 7, 657–669

Ko Swan Djien, Darwinkel-Risseeuw PS, Pilnik W (1972) Kombinierte Wirkung von Wärmebehandlung und Konservierungsmitteln auf Hefen in Traubensaft. I. Mitteilung. Versuche mit Natriumbenzoat. Confructa 17, 356–359

Ko Swan Djien, Ypma YA, Pilnik W (1973) Kombinierte Wirkung von Wärmebehandlung und Konservierungsmitteln auf Hefen in Traubensaft. II. Versuche mit Kaliumsorbat. Confructa 18, 42–44

Leimbeck R (1991) Die mikrobiologische Qualität von Salben und Cremes – Methoden, Ergebnisse, Bewertung. Parfüm Kosmet 72, 248–258

Leistner L (1978) Microbiology of ready-to-serve foods. Fleischwirtschaft 58, 2008–2011

Leistner L, Rödel W, Krispien K (1981) Microbiology of meat and meat products in high- and intermediate-moisture ranges. In: Rockland LB, Stewart GF: Water activity: Influences on food quality. Academic Press, New York, p. 855–916

Leistner L (1992) Food preservation by combined methods. Food Res Int 25, 151–158

Leistner L (1995) Principles and applications of hurdle technology. In: Gould GW: New methods of food preservation. Blackie, London

Lerche M, Linke H (1958) Versuche zur Abtötung von Salmonellen in flüssigem Eigelb mittels Sorbinsäure. Arch Lebensmittelhyg 9, 121–126
Liewen MB, Marth EH (1985) Production of mycotoxins by sorbate-resistant molds. J Food Protect 48, 156–157
Lubieniecki-von Schelhorn M (1967a) Untersuchungen über die Verteilung von Konservierungstoffen zwischen Fett und Wasser. I. Mitteilung. Physikalisch-chemische Untersuchungen. Z Lebensm Unters Forsch 131, 329–345
Lubieniecki-von Schelhorn M (1967b) Untersuchungen über die Verteilung von Konservierungstoffen zwischen Fett und Wasser. II. Mitteilung. Beziehungen zwischen physikalisch-chemischer Verteilung und antimikrobieller Wirksamkeit von Konservierungsstoffen in fetthaltigen Lebensmitteln. Z Lebensm Unters Forsch 133, 227–241
Lubieniecki-von Schelhorn M (1975) Die Sauerstoffkonzentration als bestimmender Faktor für mikrobielle Vorgänge in Lebensmitteln unter besonderer Berücksichtigung einer sauerstofffreien Verpackung. Verpack-Rundsch 26, Wiss Beilage zu Nr 1, p. 1–6
Lück E (1972) Sorbinsäure. Volume II Biochemie – Mikrobiologie. Behr, Hamburg, p. 100–105
Lück E (1973) Lebensmittel von mittlerer Feuchtigkeit. Z Lebensm Unters Forsch 153, 42–52
Lück E (1981) Schutzmaβnahmen gegen den Lebensmittelverderb durch Schimmelpilze. In: Reiβ J: Mykotoxine in Lebensmitteln. Gustav Fischer, Stuttgart, p. 437–457
Lukas E-M (1964) Zur Kenntnis der antimikrobiellen Wirkung der Sorbinsäure. II. Mitteilung: Die Wirkung der Sorbinsäure auf *Aspergillus niger* von Tieghem und andere Schimmelpilze. Zentralbl Bakteriol Parasitenkd Infektionskr Hyg II. Abt 117, 485–509
Maha M, Sudarman H, Chosdu R, Siagian EG, Nasran S (1980) Combination of potassium sorbate and irradiation treatments to extend the shelf-life of cured fish products. Comb Processes Food Irradiat Proc Int Symp p. 305–318
Ming X, Daeschel M (1993) Nisin resistance of foodborne bacteria and the specific resistance responses of *Listeria monocytogenes* Scott A. J Food Protect 56, 944–948
Parrof E (1968) Stability of methylparaben. Amer Perf Cosmet 83, 7–13
Rahn O, Conn JE (1944) Effect of increase in acidity on antiseptic efficiency. Ind Eng Chem 36, 185–187
Ray LL, Bullerman LB (1982) Preventing growth of potentially toxic molds using antifungal agents. J Food Protect 45, 953–963
Rehm H-J (1959 a) Beitrag zur Wirkung von Konservierungsmittel-Kombinationen. I. Grundlagen und Überblick zur Kombinationswirkung chemischer Konservierungsmittel. Z Lebensm Unters Forsch 110, 283–293
Rehm H-J (1959 b) Untersuchungen zur Wirkung von Konservierungsmittelkombinationen. II. Die Wirkung einfacher Konservierungsmittelkombinationen auf *Escherichia coli*. Z Lebensm Unters Forsch 110, 356–363
Rehm H-J (1960) Untersuchungen zur Wirkung von Konservierungsmittelkombinationen. IV. Wirkung einfacher Kombinationen von Konservierungsmitteln mit einigen Antibiotica auf *Escherichia coli*. Z Lebensm Unters Forsch 113, 144–152
Rehm H-J, Lukas E-M (1963) Zur Kenntnis der antimikrobiellen Wirkung der Sorbinsäure. 1. Mitteilung. Die Wirkung der undissoziierten und dissoziierten Anteile der Sorbinsäure auf Mikroorganismen. Zbl Bakteriol Parasitenkunde, Infektionskrankh Hyg II. Abt 117, 306–318
Rehm H-J, Lukas E-M, Senser F (1964) Untersuchungen zur Wirkung von Konservierungsmittelkombinationen. VIII. Die Wirkung binärer Kombinationen von Konservierungsstoffen mit weiteren Antibiotica. Z Lebensm Unters Forsch 124, 437–447
Rehm H-J, Sening E, Wittmann H, Wallnöfer P (1964) Beitrag zur Kenntnis der antimikrobiellen Wirkung der schwefligen Säure. III. Mitteilung. Aufhebung der antimikrobiellen Wirkung durch Bildung von Sulfonaten. Z Lebensm Unters Forsch 123, 425–432
Rehm H-J, Senser F, Lukas E-M (1964) Untersuchungen zur Wirkung von Konservierungsmittelkombinationen. IX. Die Wirkung binärer und trinärer Kombinationen von Konservierungsstoffen mit Nisin and Tylosin. Z Lebensm Unters Forsch 125, 258–271
Rehm H-J, Stahl U (1960) Untersuchungen zur Wirkung von Konservierungsmittelkombinationen. III. Die Wirkung einfacher Konservierungsmittelkombinationen auf *Aspergillus niger* und *Saccharomyces cerevisiae*. Z Lebensm Unters Forsch 113, 34–47

Rehm H-J, Wittmann H (1962) Beitrag zur Kenntnis der antimikrobiellen Wirkung der schwefligen Säure. I. Mitteilung. Übersicht über einflußnehmende Faktoren auf die antimikrobieller Wirkung der schwefligen Säure. Z Lebensm Unters Forsch 118, 413–429

Reiβ J (1976) Mycotoxine in Lebensmitteln. VIII. Hemmung des Schimmelpilzwachstums und der Bildung von Mycotoxinen (Aflatoxine B_1 und G_1, Patulin, Sterigmatocystin) in Weizenvollkornbrot durch Sorbinsäure und Sorboylpalmitat. Dtsch Lebensm Rundsch 72, 51–54

Robach MC (1980) Use of preservatives to control microorganisms in food. Food Technol 34: 10, 81

Rushing NB, Senn VJ (1963) The effect of benzoic, sorbic and dehydroacetic acids on the growth of citrus products spoilage organisms. Proc Fla State Hortic Soc 76, 271–276

Russell AD (1991) Mechanism of bacterial resistance to non-antibiotics: food additives and food and pharmaceutical preservatives. J Appl Bacteriol 71, 191–201

Sankaran R, Leela RK (1983) Effect of preservatives on the germination and growth of thermally injured fungal spores. Nahrung 27, 231–235

Schelhorn von M (1953) Hemmende und abtötende Wirkung von Konservierungsmitteln. Arch Mikrobiol 19, 30–44

Schelhorn von M (1954) Stand der Konservierungsmittel-Forschung. Fette, Seifen, Anstrichm 56, 221–224

Schmidt PC, Franck M (1993) Konservierung heute. Chemie in unserer Zeit 22, 39–44

Sharp TM (1976) The breakdown of food additives. An introductory survey. Scientific & Technical Surveys No. 92. The British Food Manufacturing Industries Research Association, Leatherhead

Shibasaki J, Iida S (1968) Effect of sorbic acid on the thermal injury of yeast cells. Nippon Shokuhin Kogyo Gakkai-Shi 15, 447–451

Tabata S, Kamimura H, Ibe A, Hashimoto H, Tamura Y (1994) Degradation of aflatoxins by food additives. J Food Protect 57, 42–47

Thakur B, Arya S (1993) Effect of sorbic acid on irradiation-induced sensory and chemical changes in sweetened orange juice and mango pulp. Int J Food Sci Technol 28, 371–376

Thakur B, Trehan I, Arya S (1990) Radiolytic degradation of sorbic acid in isolated systems. J Food Sci 55, 1699–1702

Tong C-H, Draughon FA (1985) Inhibition by antimicrobial food additives of ochratoxin A production by *Aspergillus sulphureus* and *Penicillium viridicatum*. Appl Environm Microbiol 49, 1407–1411

Wallhäuβer KH, Lück E (1970) Der Einfluß der Sorbinsäure auf mycotoxinbildende Pilze in Lebensmitteln. Dtsch Lebensm Rundsch 66, 88–92

York GK (1966) Effect of pimaricin on the resistance of *Saccharomyces cerevisiae* to heat, freezing and ultraviolet irradiation. Appl Microbiol 44, 451–455

Yousef AE, Marth EH (1983) Incorporation of [^{14}C] acetate into aflatoxin by resting cultures of *Aspergillus parasiticus* in the presence of antifungal agents. Eur J Appl Microbiol Biotechnol 18, 103–108

The Individual Preservatives

Table 11. Applications of the preservatives most frequently used in conjunction with the main groups of foodstuffs

	Nitrate, nitrite	Sulfur dioxide	Sucrose	Hexa-methylene tetramine	Formic acid	Acetic acid	Propionic acid	Sorbic acid	Benzoic acid	*p*-Hydroxy-benzoic acid esters	Biphenyl, *o*-Phenyl, phenol, Thiabend-azole	Smoke
Fat emulsions	–	–	–	–	–	–	–	++	+	–	–	–
Cheese	(+)	–	–	(+)	–	–	+	++	(+)	(+)	–	+
Meat products	++	(+)	–	–	–	–	–	+	–	(+)	–	++
Fish products	+	–	–	(+)	–	++	–	+	+	(+)	–	++
Vegetable products	–	+	(+)	–	(+)	++	–	++	++	–	–	–
Fruit products	–	++	++	–	(+)	+	–	++	++	–	(+)	–
Soft drinks	–	++	++	–	(+)	–	–	++	++	–	–	
Wine	–	++	–	–	–	–	–	++	–	–	–	–
Baked goods	–	–	++	–	–	–	++	++	–	–	–	–
Confectionery	–	–	++	–	–	–	–	++	(+)	(+)	–	–

Key of symbols:
++ used frequently
\+ used occasionally
(+) used in exceptional cases only
– not used

Chapter 6

Common Salt

6.1 Synonyms

IUPAC: Sodium chloride
English: sodium chloride, table salt, “salt”. *German:* Speisesalz, “Salz”. *French:* Chlorure de sodium, sel de cuisine, “sel”. *Italian:* Cloruro di sodio, sale de cucina, “sale”. *Spanish:* Cloruro sódico, sal común, sal de cocina, “sal”. *Russian:* Хлорид натрия, иоваренная соль, “соль”.

6.2 History

Salt was even mentioned in the Old Testament as a food additive with a ritual character. Its use as a food preservative was well known in Ancient Egypt (Netolitzky 1913), the Middle East and Ancient Rome (Bergier 1989, Forbes 1965, Schleiden 1875). Muria, garum, liquamen and alec (or alex) were all preparations containing common salt as a main ingredient and employed to flavor and preserve foods of various kinds (Binkerd and Kolari 1975). In central and northern Europe salt was an important trading commodity in the Middle Ages, as recalled by the names of towns such as Halle. Although twice as expensive as beef two hundred years ago, salt remained for centuries the most important if not the only preservative to be employed on any large scale, especially for meat, fish and vegetables.

Salt has retained its importance in food preservation to the present day, although it is now used less as a preservative in its own right than in combination with other preservatives and preservation methods.

6.3 Commercially Available Forms

Common salt is sold in various grain sizes. Coarser or finer varieties afford particular advantages, depending on the intended use. A distinction is drawn between sea salt, rock salt and evaporated granulated salt according to origin. These different types may contain varying minor quantities of other mineral constituents.

6.4 Properties

NaCl, molar mass 58.44, colorless, odorless, salty-flavored, cubic crystals which melt at 801 °C. At room temperature, 100 g water will dissolve about 36 g NaCl. A saturated common salt solution at room temperature contains 26.5 g NaCl per 100 g or 31.1 g per 100 ml solution. Its pH value is 6.7 to 7.3 and its density 1.2.

6.5 Analysis

To determine the common salt content of a foodstuff, the foodstuff itself or its ash is extracted with warm water and the usual technique employed to determine the sodium content, e.g. by flame photometry, or the chloride content, e.g. by titration using Mohr's or Volhard's method.

6.6 Production

Common salt is obtained from rock salt deposits or seawater. As a rule, the rock salt obtained by mining is not sufficiently pure for use in food. To produce salt for culinary purposes, rock salt is dissolved underground in water and, after appropriate purification, is dried by evaporation in large pans. To obtain sea salt, seawater is allowed to evaporate in shallow tanks in hot countries by solar heat, thus causing the individual salts contained in the seawater to crystallize out in succession (Kaufmann 1960).

6.7 Health Aspects

6.7.1 Acute Toxicity

The LD50 of common salt was determined by peroral administration of the salt in concentrated aqueous solutions to fasted rats as 3.75 g/kg body weight (Boyd and Shanas 1963, Boyd and Boyd 1973). For humans a dosage of 35–40 g common salt is highly toxic (Meneely and Batterbee 1976). There have been repeated cases of acute poisoning, for example where common salt is mistaken for sugar (Seeger 1994).

6.7.2 Subchronic Toxicity

When common salt is administered in concentrated aqueous solutions to fasted rats over 100 days, i.e. over about one-tenth of their life span, the LD50 is 2.7 g/kg body weight. The equivalent figure for non-fasted rats is 6.14 g/kg body weight, the difference being attributable to the dilution of the salt in the feed and the delay in absorption (Boyd et al. 1966, Boyd and Boyd 1973). When common salt is added to the feed, substantially higher concentrations are tolerated (Meneely et al. 1952, Batterbee and Meneely 1978). Common salt has neither a teratogenic effect (Food and Drug Research Laboratories 1974) nor a mutagenic effect on *Saccharomyces cerevisiae or Salmonella typhimurium* (Litton Bionetics 1976).

6.7.3 Chronic Toxicity

A quantity of 2.8 to 5.6% common salt, relative to the feed, retards growth and shortens the life span (Meneely et al. 1953).

6.7.4 Biochemical Behavior

Common salt is essential for maintaining the osmotic pressure of the body fluids and vital as a source of sodium. The physico-chemical action is capable of producing toxic effects due to excessive concentrations; these extend primarily to the digestive tract but secondarily to virtually all other organs of the body (Boyd and Boyd 1973, Federation of American Societies for Experimental Biology 1979). Owing to its good water solubility, common salt is excreted relatively quickly. In humans, common salt is contraindicated for certain diseases of the heart, circulatory system and kidneys. Salt substitutes are used in such instances, but these have no importance as food preservatives.

6.8 Regulatory Status

As a vital dietary constituent which has been used in food technology for centuries, common salt is subject to scarcely any legal restrictions, even so far as maximum permissible concentrations in foods are concerned. Exceptions to this are dietary foods labelled as "low salt content" or "salt-free". In most countries, common salt is not defined as an additive.

6.9 Antimicrobial Action

6.9.1 General Criteria of Action

Common salt lowers the water activity (a_w value) of a system and thus renders conditions less favorable to microbial life. Its mode of action is therefore comparable with that of drying; hence the term "chemical drying" to describe the use of common salt. However, since the aw value of saturated common salt solution is only in the region of 0.75 and a number of microorganism varieties are able to grow even

Table 12. Water activity of common salt solutions (Robinson and Stokes 1959)

a_w value	Content of solution in g NaCl/100 g H_2O
0.995	0.88
0.99	1.75
0.98	3.57
0.96	7.01
0.95	8.82
0.94	10.34
0.92	13.50
0.90	16.54
0.88	19.40
0.86	22.21
0.85	23.55
0.84	24.19
0.82	27.29
0.80	30.10
0.78	32.55
0.76	35.06
0.75	36.06

below this limit, it is impossible to protect a foodstuff reliably from all microbial attack by using common salt alone, quite apart from the virtually unacceptable restrictions imposed on taste (Kushner 1971).

The foods to be preserved can be immersed in solutions containing greater or smaller amounts of common salt (brines). Alternatively, dry common salt can be added to the food. The resulting osmotic removal of water from the food adjusts the water activity to a certain level according to the quantity of common salt added. Table 12 gives a survey of the relationships involved.

According to their salt tolerance, microorganisms are defined as slightly halophilic (salt-tolerant), moderately halophilic or extremely halophilic. The first-named grow best in the presence of some 1–5% common salt. Moderately halophilic microorganisms tolerate 5–20% common salt in the medium, and extremely halophilic strains up to 30%.

The effect of common salt in lowering the water activity does not in itself adequately explain its antimicrobial action. Thus certain clostridia strains, for example, grow in the presence of common salt only if the water activity is 0.96 or more, but in the presence of glycerin they continue to grow even if the water activity is as low as 0.93 (Baird-Parker and Freame 1976). In addition, bacteria are more inclined to accumulate certain amino acids when the water activity is lowered, and this inhibits their growth (Sinskey 1980). Lastly, common salt reduces the solubility of oxygen in water. Hence, the quantity of oxygen available to aerobic microorganisms in products containing high concentrations of common salt is only a fraction of that in substances with a low salt content (Lubieniecki-von Schelhorn 1975).

From concentrations of as little as 2% upwards, common salt intensifies the action of preservatives in the narrow sense of the term (von Schelhorn 1951). The minimum inhibitory concentration of sorbic acid for yeasts and molds in the presence of 4 to 6% common salt is between one-half and one-third, and in the presence of 8% common salt about one-quarter the concentration of sorbic acid used on its own (Lück 1972). This effect is especially marked in the acid pH range in relation to yeasts (Smittle 1977) and clostridia (Baird-Parker and Freame 1967). The combination of common salt with physical methods of preservation, especially refrigeration and drying, is also of considerable practical importance (Sofos 1983, Barbuti et al. 1989, Papageorgiou and Marth 1989). Common salt increases the heat resistance of molds (Doyle and Marth 1975) and bacteria (Bean and Roberts 1975) as a result of osmotic effects. It is documented as having the reverse effect on clostridia (Hutton et al. 1991).

With a number of products a phenomenon closely related to the salting of foods is that of natural pickling. In this the addition of salt initiates a microflora selection process which favors bacteria that form lactic acid.

The direct enzyme-inhibiting action of common salt is of little practical consequence in explaining its antimicrobial action. Indeed, there are some enzymes whose activity is actually increased by low concentrations of common salt.

6.9.2 Spectrum of Action

As common salt acts chiefly by reducing the water activity, its spectrum of action is governed by the demands imposed on the water activity by the various microorganisms (Ingram and Kitchell 1967). The limiting values of the water activity for some important microorganisms occurring in foods are shown in table 10. The microorganisms that tolerate relatively high salt concentrations, mention should be made of Torulopsis and Torula yeasts, oospora, various staphylococci and lactic acid bacteria.

Some microorganisms are halophilic in the true sense, i.e. they not only tolerate high common salt concentrations but actually grow better in the presence of common salt. Bacteria of this type are, however, of little importance as food spoilants.

6.10 Fields of Use

6.10.1 Fat Products

Of these, only emulsified fats are susceptible to microbial attack, e.g. butter and margarine. Common salt is important as a preservative for both.

Some 0.3 to 2% of common salt, relative to the total weight, is added in dry or dissolved form to butter after the butter grains are washed and before kneading commences. This represents an addition of some 1 to 13% common salt in the water phase, which is the phase susceptible to microbial attack.

In margarine the quantity of common salt employed is up to 3%, i.e. up to 19% relative to the water phase to which it is added.

6.10.2
Dairy Produce

Common salt is highly important as a preservative for cheese. Depending on the type of cheese, the salt is added either in dry form to the cheese curd and/or to the cheese surface (dry salting) or – more commonly – in the form of salt solutions (brine baths). However, these virtually saturated common salt solutions with a pH value of 5.2 readily become populated by yeasts and salt-tolerant bacteria which may constitute sources of infection for the cheese. According to their types and size, the cheeses remain in the salt solutions for between one hour (Camembert) and five days (Emmentaler). Depending on its type, ripe cheese has a salt content between 1 and 3%; only in exceptional cases is it higher or lower. A common salt content of 5%, relative to the water content of the cheese, may be regarded as optimum. When the salt is added to the cheese curd, a relatively fine grain size of up to 1 mm is preferred. The ideal grain size for dry salting is 1.8 to 2 mm.

Cheeses that take relatively long to ripen are washed in common salt solution at intervals to suppress surface mold. However, neither this treatment nor any other method of salting cheese is sufficient to protect it completely from microbial attack, especially mold growth; so ideally the salt – which in cheese is just as much a flavor additive as a preservative – should be supplemented by true mold prevention agents such as sorbic acid.

6.10.3
Egg Products

Occasionally, common salt is still employed in concentrations of 5 to 8% for preserving liquid whole egg and liquid egg yolk. Pickled eggs are hard-boiled eggs preserved by immersion in relatively highly concentrated salt solutions.

6.10.4
Meat Products

Salted meat played a major part as a staple food for centuries. Today, common salt is used on a substantial scale as an ingredient of pickling brines or as a preservative in conjunction with other preserving processes such as refrigeration, drying and smoking. In such instances common salt displays a good antimicrobial effect in concentrations of as little as 1–3%. These relatively small additions reduce the water activity sufficiently to suppress the growth of many important putrefactive bacteria, e.g. in sausage products, ham and other pickled meats.

6.10.5
Fish Products

The salting of fish is an extremely old method of preservation. In the case of fish products, common salt has retained its great importance as a preservative to the present day.

Salt is used chiefly to preserve herrings although also for anchovies, sprats, cod, salmon and fish roes. Since the fourteenth century (Willem Beukelzoon, Biervliet, Netherlands), herrings have been "cleaned" or "gutted" before being salted, i.e. their inner organs, apart from the milts and roe, are removed. The salt is added on board the fishing craft or after the catch is landed. It may be in solid form (dry salting) or brine (brine salting). In dry salting, the grain size of the salt is of some importance. Unduly fine salt will penetrate the fish meat rapidly and the fish will deteriorate outwards from the inside. If the salt employed is too coarse, there will be a risk of uneven distribution.

A distinction is drawn between various degrees of salting. An addition of 0.5 to 2% common salt, relative to the weight of fish, may be made at sea, this representing 0.7 to 3% in the water of the fish tissue. In this concentration, common salt alone does not provide adequate protection from microbial spoilage. The fish must also be refrigerated and the preserving time needs to be kept very short.

Slight salting connotes the addition of about 10% common salt in dry or dissolved form, relative to the weight of the fish, and at all events less than 20%, relative to the water in the fish tissue. A typical product manufactured by slight salting is Dutch or German matie herring. However, even the salt concentration used in this instance provides only limited keeping power against microbial attack.

Moderate salting likewise implies an addition of less than 20% salt, relative to the water in the fish tissue, but a higher salt concentration than in slight salting. The common salt solutions used in brine salting are of 15 to 18%.

Hard salting involves the addition of common salt in concentrations of more than 14%, relative to the fish meat, or 20 to 24%, relative to the water in the fish tissue. Heavily salted fish are a starting material for further processing and resist microbial spoilage for considerably longer. However, the growth of salt-tolerant or halophilic microorganisms is not impossible in these products. To enhance their keeping power still further, the fish can also be dried, as practiced, for instance, in the production of klipfish.

Common salt is also important as a preservative for fish roes and caviar (salt content 3–10%), anchovies, fish pastes (salt content up to 20%) and fish preserves (salt content 2–5%). For various reasons, common salt in this case is rarely employed as a preservative on its own but used in conjunction with other preserving substances such as smoke, boric acid (caviar), saltpeter (anchovies), acetic acid, hexamethylene tetramine, benzoic acid and sorbic acid (marinades), as well as oil (salmon substitute in oil).

6.10.6
Vegetable Products

Common salt is used alone as a preservative in what are known on the European mainland as "salt vegetables". These are intermediate products intended for further industrial processing, the main vegetables preserved in this way being asparagus, beans, cabbage, carrots, turnips, pearl onions, mushrooms and olives. These are placed in common salt solution of 15 to 25%, according to the vegetables concerned. Owing to the high salt concentration, virtually no lactic acid fermentation takes place but the occurrence of film yeasts is not impossible.

Vegetables pickled in weak solutions of salt and subjected to lactic acid fermentation, e.g. sauerkraut, pickled gherkins and olives, are not included under the term "salt vegetables". The main preservative in this instance is not common salt but the organic acids added or formed by the fermentation.

6.10.7 Fruit Products

Owing to its flavor, common salt is scarcely used at all as a preservative in fruit products. The only special product worth mention is the raw material (Cederfrüchte) used in the manufacture of succades. In this application common salt is employed as a preservative in the form of sea water (Livorneser Purgierung) or as a 6 to 8% brine in an intermediate stage before the products are finally preserved with sugar.

6.11 Other Effects

Besides its preservative action, common salt has a considerable number of other effects, the majority of which are not undesirable. The property of salt as a flavor enhancer should be mentioned first, as in many foods this, rather than a preservative action, is its main function. The concentrations of common salt required for flavor enhancement are generally much lower than those needed for preservation purposes; so in principle foods preserved with common salt alone are rarely suitable for direct consumption. Either they are raw materials for further industrial processing or else they need to be desalted, e.g. by immersion in water.

Common salt has a variety of influences on proteins. In view of the high applied concentrations of common salt as a preservative, this is not surprising. These influences include the swelling effect on meat, which affects its water-binding capacity, and the properties of common salt in making fish palatable.

On the whole, foods preserved with common salt have a greater tendency to oxidize, i.e. especially for the fat constituent to turn rancid. The cause of this is the common salt itself, although traces of heavy metals in the salt may also tend to promote oxidation. The action of common salt in encouraging oxidation is of practical importance chiefly in meat and fish products.

Finally, mention should be made of the fact the common salt removes water-soluble ingredients such as minerals, vitamins and proteins from the food by the osmotic extraction of water. Consequently the biological nutrient value of foods preserved with salt is generally lower than that of the corresponding fresh products.

6.12 Literature

Baird-Parker AC, Freame B (1967) Combined effect of water activity, pH and temperature on the growth of Clostridium botulinum from spore and vegetative cell inocula. J Appl Bacteriol 30, 420–429

Barbuti S, Ghhisi M, Companini M (1989) Listeria in meat products. Isolation, incidence and growth characteristics. Ind Conserve 64, 221–224
Batterbee HD, Meneely GR (1978) The toxicity of salt. Crit Rev Toxicol 5, 355–376
Bean PG, Roberts TA (1975) Effect of sodium chloride and sodium nitrite on the heat resistance of Staphylococcus aureus NCTS 10652 in buffer and meat macerate. J Food Technol 10, 327–332
Bergier JF (1989) Die Geschichte vom Salz. Campus, Frankfurt
Binkerd EF, Kolari OE (1975) The history and use of nitrate and nitrite in the curing of meat. Food Cosmet Toxicol 13, 655–661
Boyd EM, Abel MM, Knight LM (1966) The chronic oral toxicity of sodium chloride at the range of the LD50 (0.1 L). Can J Physiol Pharmacol 44, 157–172
Boyd EM, Boyd CE (1973) Toxicity of pure foods. CRC Press, Cleveland
Boyd EM, Shanas MN (1963) The acute oral toxicity of sodium chloride. Arch Intern Pharmacodyn Ther 144, 86–98
Doyle MP, Marth EH (1975) Thermal inactivation of conidia from Aspergillus flavus and Aspergillus parasiticus. II. Effects of pH and buffers, glucose, sucrose and sodium chloride. J Milk Food Technol 38, 750–758
Federation of American Societies for Experimental Biology (1979) Evaluation of the health aspects of sodium chloride and potassium chloride as food ingredients. PB-298139. National Technical Information Service. US Department of Commerce, Springfield
Food and Drug Research Laboratories (1974) Teratologic evaluation of FDA 71–70 (sodium chloride) in mice, rats and rabbits. PB-234878. National Technical Information Service. US Department of Commerce, Springfield
Forbes RJ (1965) Studies in ancient technology, Volume 3. Brill, Leiden, p. 164–181
Hutton MT, Koskinen MA, Hanlin JH (1991) Interacting effects of pH and NaCl on heat resistance of bacterial spores. J Food Sci 56, 821–822
Ingram M, Kitchell AG (1967) Salt as a preservative for foods. J Food Technol 2, 1–15
Kaufmann DW (1960) Sodium chloride. The production and properties of salt and brine. Reinhold, New York
Kushner DJ (1971) Influence of solutes and ions on microorganisms. In: Hugo WB (1976) Inhibition and destruction of the microbial cell. Academic Press, London, p. 259–282
Litton Bionetics (1976) Mutagenic evaluation of compound. FDA 71–70.007647–14–5, sodium chloride reagent, ACS. PB-257870 National Technical Information Service. US Department of Commerce, Springfield
Lubieniecki-von Schelhorn M (1975) Die Sauerstoffkonzentration als bestimmender Faktor für mikrobielle Vorgänge in Lebensmitteln unter besonderer Berücksichtigung einer sauerstofffreien Verpackung. Verpack-Rundsch 26, Wiss Beilage zu Nr 1, p. 1–6
Lück E (1972) Sorbinsäure. Chemie-Biochemie-Mikrobiologie-Technologie-Recht, Volume 2. Behr, Hamburg, p. 91
Meneely GR, Batterbee HD (1976) Sodium and potassium. Nutr Rev 34, 225–235
Meneely GR, Tucker RG, Darby WJ (1952) Chronic sodium chloride toxicity in albino rats. J Nutr 48, 489–498
Meneely GR, Tucker RG, Darby WJ, Auerbach SH (1953) Chronic sodium chloride toxicity: Hypertension, renal and vascular lesions. Ann Int Med 39, 991–998
Netolitzky F (1913) Nahrungs- und Heilmittel der Urägypter. Z Unters Nahrungs Genubm 26, 425–427
Papageorgiou DK, Marth EH (1989) Behaviour of Listeria monocytogenes at 4 and 22° C in whey and skim milk containing 6 or 12% sodium chloride. J Food Protect 52, 625–630
Robinson RA, Stokes RH (1959) Electrolyte solutions. The measurement and interpretation of conductance, chemical potential and diffusion in solutions of simple electrolytes, 2nd edn. Butterworths, London
Schelhorn M von (1951) Wirksamkeit und Wirkungsbereich chemischer Konservierungsmittel für Lebensmittel. Z Lebensm Unters Forsch 92, 256–266
Schleiden MJ (1875) Das Salz. Seine Geschichte, seine Symbolik und seine Bedeutung im Menschenleben. Engelmann, Leipzig

Seeger R (1994) Giftlexikon: Natrium. Dtsch Apotheker-Ztg 134, 105–115
Sinskey AJ (1980) Mode of action and effective application. In: Tilbury RH: Developments in food preservatives – 1. Applied Science Publishers, London, p. 113–115
Smittle RB (1977) Influence of pH and NaCl on the growth of yeasts isolated from high acid food products. J Food Sci 42, 1552–1553
Sofos JN (1983) Antimicrobial effects of sodium and other ions in foods. A review. J Food Safety 6, 45–78

Chapter 7

Carbon Dioxide

7.1 Synonyms

IUPAC: Carbon dioxide
German: Kohlendioxid, Kohlensäureanhydrid, "Kohlensäure". *French:* Bioxyde de carbone, gaz carbonique, "acide carbonique". *Italian:* Biossido di carbonio, anidride carbonica, "acido carbonico". *Spanish:* Dióxido de carbono, anhidrido carbónico, "ácido carbónico". *Russian:* Двуокиь углерода, карбонангидрид, "угольная кислота".

7.2 History

Since ancient times the carbon dioxide remaining from alcoholic fermentation in drinks or from the respiration of stored cereals has been exploited, largely unwittingly, as a food preservative. The controlled addition of carbon dioxide from other sources is, however, a recent development.

7.3 Commercially Available Forms

Carbon dioxide is generally employed in the form of liquefied gas, which can be easily and reliably metered. Solid carbon dioxide (dry ice) is highly important as a refrigerant, but that application lies outside the scope of this review. Its use as a preservative against microbial spoilage in the narrower sense of the term is of minor significance only.

7.4 Properties

CO_2, molar mass 44.21, non-combustible gas, colorless at room temperature and normal pressure, with an acidic odor and taste. The density of carbon dioxide is about 1.5 times that of air. At 0 °C carbon dioxide can be compressed under a pressure of 34.85 bar into a colorless liquid; at 20 °C, 55.4 bar is required. The water-solubility of carbon dioxide at room temperature is 1 liter per liter and thus higher than that of many other gases. As a result, when protective gas mixtures containing carbon dioxide are used, varying amounts of carbon dioxide may be absorbed by the foods, depending on the pH value, storage temperature, water activity and other parameters, thereby causing the concentration in the gas atmosphere to change (Löwenadler and Rönner 1994).

7.5 Analysis

For quantitative determination the carbon dioxide emitted by the substance under investigation can be passed into barium hydroxide solution. The quantity of carbon dioxide can then be ascertained from the amount of barium carbonate obtained by titration. A more exact method is that of volumetric determination, in which carbon dioxide is first bound as a carbonate by the addition of sodium hydroxide. The carbon dioxide then driven off again by sulfuric acid can be directly determined volumetrically or by coulometry (Löwenadler and Rönner 1994). Carbon dioxide can also be detected by a method using special test tubes (Drägerwerk 1991).

7.6 Production

Carbon dioxide can be obtained in highly pure form from natural sources. It is also frequently produced from the gas mixture deriving from the burning of coke. This is fed by way of absorption towers in which the carbon dioxide is bound in the form of bicarbonate.

7.7 Health Aspects

Figures on the concentrations of carbon dioxide that rapidly produce death in animals when present in respiratory air vary between 30% by volume and 60% by volume (in the presence of 20% by volume oxygen). When inhaled over a relatively long period of time, concentrations of carbon dioxide exceeding 3% by volume in the air breathed may be dangerous to the subject. The maximum allowable workplace concentration for carbon dioxide is 5000 mg/m^3.

7.8 Regulatory Status

In most countries carbon dioxide (E 290) is subject to virtually no food law limitations. In some countries, the requirement exists that bottled mineral waters and effervescent drinks prepared from such mineral waters must have the same carbon dioxide content as the natural starting material. As far as wine is concerned, there are some countries in which the carbon dioxide must not exceed certain maximum levels in order to maintain a distinction with sparkling wine.

7.9 Antimicrobial Action

7.9.1 General Criteria of Action

The effect of carbon dioxide is based on several factors. Firstly, carbon dioxide displaces oxygen, which is a vital requirement for many microorganisms. Secondly, it has an antimicrobial action of its own in that it intervenes in the respiratory metabolism of various microorganisms. Finally, if present in relatively large quantities, carbon dioxide may change the pH conditions on the surface of the food and thus deprive some microorganisms of the conditions vital to life (Dixon and Kell 1989; Haas et al. 1989).

The technological process of storing foods in atmospheres modified with inert gases is known today as modified-atmosphere packaging (MAP). Since the main action of carbon dioxide is to displace oxygen, carbon dioxide obviously has little or no effect in inhibiting the growth of anaerobic microbes. Foods preserved by MAP technology are thus not entirely safe since carbon dioxide inhibits only aerobic putrefactants, whose presence is apparent to consumers as actual spoilage. The carbon dioxide has little effect on the growth of pathogenic anaerobic organisms, which may cause infections or poisoning (Farber 1991). The growth of *Campylobacter jejuni* has even been described as being stimulated by MAP (Hanninen et al. 1984). The same applies to *Clostridium sporogenes* (Zee 1984, Dixon and Kell 1989). The formation of toxins by *Clostridium botulinum* is assisted by relatively low CO_2 concentrations, whereas high CO_2 concentrations (45–70%) have an inhibitory effect (Lambert et al. 1991).

Other disadvantages of the process are that it requires a controlled temperature (cold storage) and the selection, for each food, of a defined gas composition which has to be determined experimentally. It is also a cost-intensive process which can be carried out only by trained personnel. Against these drawbacks there is the advantage that, depending on the storage temperature, foods placed in MAP can be kept for much longer than foods stored conventionally.

Figure 7 shows the growth of spoilage bacteria on pork at different temperatures in air and CO_2 atmospheres (CO_2 in the gas phase: 100%). The time is stated in days and the measurement parameter used is 10^6 CFU/cm^2 meat surface (CFU = colony-forming unit) (Gill and Harrison 1989).

The use of atmospheres with high carbon dioxide partial pressure is of particular interest for foods with low water activity.

Carbon dioxide does not generally eliminate the microorganisms, at least not within short periods, but merely inhibits their growth. The microorganisms are maintained in a latent phase, in which they cease to reproduce but are capable of certain enzymatic reactions. A rapid microbicidal effect can be achieved by high carbon dioxide pressures, possibly in combination with low temperatures. This technology is becoming increasingly important in the protection of stored goods by the food industry, e.g. silo storage of flour, tea, or spices. Pressures of between 10 and 30 bar and exposure times of 30–240 min are used or, alternatively, normal

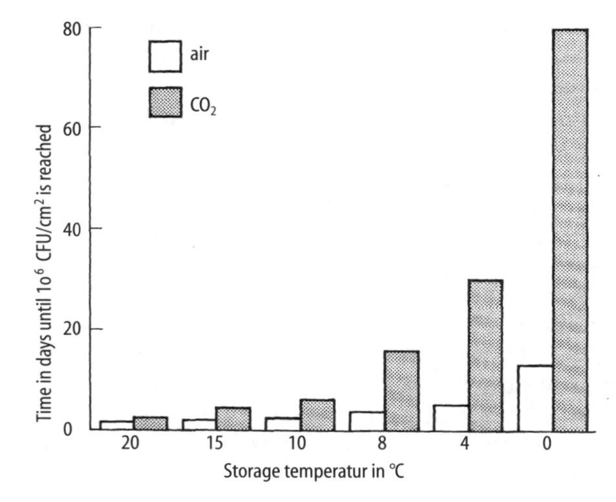

Fig. 7. Growth of spoilage bacteria on pork

pressure (CO_2 > 60 %) with exposure times of 10–20 days (Rau 1993). Extremely high pressures (several 1000 bar) are also suitable for the destruction of insects in rice or other raw materials, though less expensive alternatives at lower pressures have recently been described (Locatelli and Daolio 1993). In addition, extremely high pressures are used to inactivate pectinesterases in fruit juice (Buchheim and Prokopek 1992).

Packages in which carbon dioxide is used as a protective gas need to be made from packing materials with the lowest possible gas permeability, such as polyester/polyethylene or polyester/aluminium/polyethylene composite films (Barnett et al. 1987). Any decision whether to use protective gas packaging or its competitor, namely vacuum packaging, is generally governed by extraneous factors and depends essentially on the configuration and nature of the package contents, as well as on the flexibility and strength of the pack.

7.9.2
Spectrum of Action

Carbon dioxide acts mainly against obligate aerobic microorganisms. Molds are highly resistant to carbon dioxide and their growth cannot be completely halted by treatment with carbon dioxide gas at normal pressure (Lubieniecki-von Schelhorn 1974). This applies especially at temperatures around 20 °C (Eklund and Jarmund 1983). The preservative effect is increased only if the oxygen content is reduced below 0.2% (Lubieniecki-von Schelhorn 1974). When the carbon dioxide present is between 10 and 90%, relative to the storage atmosphere, aflatoxin-forming microorganisms produce a smaller quantity of toxin the higher the carbon dioxide content of the ambient atmosphere (Shih and Marth 1973).

Yeasts are likewise only very slightly sensitive to carbon dioxide. Under the action of carbon dioxide there are some circumstances in which yeasts are capable of certain metabolic reactions that cause food spoilage. A number of yeasts, for instance, can ferment sugar to alcohol even under pressurized carbon dioxide and without reproducing. Surface yeasts are kept under control to a certain extent by carbon dioxide. The action of carbon dioxide against yeasts increases with the carbon dioxide concentration.

The behavior of the different bacteria strains towards carbon dioxide is extremely varied. Gram-negative psychrophilic strains, pseudomonades (Eklund and Jarmund 1983) and *Achromobacter* strains and *Escherichia coli* (Molin 1983) are all particularly sensitive. The inhibitory effect of carbon dioxide on salmonella is especially pronounced in the presence of potassium sorbate. The antimicrobial activity of carbon dioxide high-pressure processes can also be used against salmonella or listeria (Wei et al. 1991). In contrast, many lactic acid bacteria (Eklund and Jarmund 1983, Molin 1983) and clostridia (Dixon and Kell 1989, Faber 1991) are fairly resistant to carbon dioxide. In certain circumstances carbon dioxide may even promote the growth of certain salmonella, lactic acid bacteria and *Campylobacter* strains (Farber 1991, Church and Parsons 1995).

7.10 Fields of Use

7.10.1 Dairy Produce

For cheese in consumer packs, carbon dioxide is employed as an inert protective gas against oxidation and microbiological spoilage. Carbon dioxide suppresses the growth of yeasts, molds and cold-resistant bacteria, in both fresh and hard cheeses (Kosikowski and Braun 1973). The carbon dioxide concentration should be 100%, relative to the storage atmosphere. Mixtures of carbon dioxide and nitrogen are also very effective.

7.10.2 Meat Products

If fresh meat is kept in cold storage in a 15 to 40% carbon dioxide atmosphere, this will substantially increase its storage life. The preserving effect will increase with a rise in the carbon dioxide concentration (Gill and Molin 1989). The carbon dioxide acts primarily on molds, pseudomonades and achromobacter; yeasts, lactobacilli and *Microbacterium thermosphactum* are less sensitive. Combinations of CO_2 with other protective gases are also very effective (Holley et al. 1994).

Higher contents of carbon dioxide in the storage atmosphere may cause changes in the color and flavor of the meat, especially if the oxygen content is below a certain limit. This is described as occurring, for example, with chilled sliced roast beef (Penney et al. 1993). The literature, however, is not consistent so far as optimum concentrations of carbon dioxide and oxygen are concerned. Whereas a number of authors give a maximum of 25% and a minimum of 5% as the limiting values for carbon dioxide and oxygen respectively, others consider 70 to 85% for oxygen and 15% for carbon dioxide to be optimum values (see Table 13).

For frankfurter-type sausages, a 50–80% carbon dioxide content in the inert gas atmosphere is described as giving an optimum antimicrobial effect against *Listeria monocytogenes* (Krämer and Baumgart 1992).

7.10.3 Marine Animals

The storage life of fish, e.g. mullet, can also be extended by storage in inert gas atmospheres (Gupta et al. 1991). In addition, carbon dioxide combined with heat can inhibit polyphenol oxidases which cause undesired browning in spiny lobsters (Chen et al. 1993).

7.10.4 Fruit and Vegetables

Inert gas atmospheres are a relatively harmless way of suppressing changes caused by microorganisms and enzymes in fruit and some vegetables during storage, e.g. in potatoes (Mazza and Siemens 1990), figs (Colelli et al. 1991), pears (Garcia and Streif 1993), and certain varieties of mushrooms, including Shiitake mushrooms (Kim et al. 1989).

7.10.5 Drinks

The storage of fruit juices under compressed carbon dioxide is a frequently practised method of preservation. This method was invented by a Swiss, Böhi. The process is known as the Seitz-Böhi method in the fruit juice industry.

To ensure that the juices keep fresh, carbon dioxide concentrations of 1.5% are necessary, this corresponding to a gauge pressure of some 7 bar at 15 °C. The microorganisms, especially the yeasts, are not killed by the carbon dioxide even under the influence of pressure. Their growth is merely inhibited and they remain perfectly capable of certain enzymatic reactions; for instance, yeasts retain some ability to produce alcohol. Carbon dioxide has only a limited action against lactic acid fermentation, so the risk of lactic acid fermentation exists even if the juices are stored under compressed carbon dioxide. Where permissible the action of carbon dioxide can be improved by adding small quantities of sulfur dioxide.

Carbon dioxide is also very important as a preservative in soft drinks. In these, the carbon dioxide concentration, at a pressure of 2 to 4 bar, is far lower than in the Seitz-Böhi method. Consequently, the storage life attainable is shorter but still generally sufficient for practical purposes. Experience has long shown that drinks containing carbon dioxide have better resistance to microbial spoilage than do still drinks. The microorganism content of such drinks declines with increasing storage time and rising carbon dioxide concentration.

Finally, the antimicrobial action of carbon dioxide is utilized in the production of wine. By fermenting the must in pressure tanks, it is possible to control the fermentation and stop the process at a certain alcohol/sugar content so that the wine obtained has a certain residual sweetness (Troost 1988). Wine can be directly impregnated with carbon dioxide to protect it from microbial spoilage such as secondary fermentation. A good effect is obtained with 0.6 to 1.2 g carbon dioxide per liter wine. Mixtures of carbon dioxide and nitrogen can also be used to good effect.

7.10.6 Baked Goods

The storage and packaging of part-baked doughs, cakes or cut bread in inert gas atmospheres is now a widely used technological process, though subject to limitations because of the large quantity of air and oxygen present in baked goods. In practice the packs and the bread need to be evacuated and/or flushed with carbon dioxide. The carbon dioxide concentrations in such systems vary according to the conditions (a_w value, storage temperature and microorganisms), ranging from 62 to 99% (Avital et al. 1990). The storage life of baked goods can also be improved by combining carbon dioxide with other antimicrobial substances, such as ethanol (Vora et al. 1987) or sorbic acid (Smith et al. 1988).

7.11 Other Effects

As it has only a low level of reactivity and acts by driving off oxygen, carbon dioxide exercises what is in a certain sense an "antioxidative" effect. It is employed, for instance, in the packaging of fat products, dairy produce, snacks, peanuts and other dry products sensitive to oxidation. The "antioxidative" effect of carbon dioxide is often far more important than its antimicrobial action. However, like the influences of carbon dioxide on the respiration of stored fruit and vegetables (Kubo et al. 1990), it lies outside the scope of this chapter. In drinks, the addition of carbon dioxide generally has a beneficial effect on the organoleptic properties. Additions of carbon dioxide enhance the refreshing effect of many drinks. Finally, mention may be made of the hyperemic effect of carbon dioxide, which accelerates the absorption of other substances by the mucous membrane of the stomach. This phenomenon influences, for example, the accelerated absorption of alcohol from champagne or sparkling wine.

7.12 Literature

Avital Y, Mannheim C, Miltz J (1990) Effects of carbon dioxide atmosphere on staling and water relations in bread. J Food Sci 55, 413–416, 461

Barnett H, Conrad J, Nelson R (1987) Use of laminated high and low density polyethylene flexible packaging to store trout (*Salmo gairdneri*) in a modified atmosphere. J Food Protect 50, 645–651

Buchheim W, Prokopek D (1992) Die Hochdruckbehandlung. Dtsch Milchwirtsch 43, 1374–1378

Chen J, Balaban M, Cheng Wei I, Gleeson R, Marshall M (1993) Effects of carbon dioxide on the inactivation of Florida spiny lobster polyphenol oxidase. J Sci Food Agric 61, 253–259

Church I, Parsons A (1995) Modified atmosphere packaging technology: a review. J Sci Food Agric 67, 143–152

Colelli G, Mitchel F, Kader A (1991) Extension of postharvest life of mission figs by carbon dioxide enriched atmospheres. Hort Sci 26, 1193–1195
Dixon N, Kell D (1989) The inhibition by carbon dioxide on the growth and metabolism of microorganisms. J Appl Bacteriol 67, 109–136
Drägerwerk (1991) Dräger-Röhrchen Handbuch. Drägerwerk, AG Lübeck
Eklund T, Jarmund T (1983) Microculture model studies on the effect of various gas atmospheres on microbial growth at different temperatures. J Appl Bacteriol 55, 119–125
Elliott P, Gray R (1981) Salmonella sensitivity in a sorbate/modified atmosphere combination system. J Food Protect 44, 903–908
Farber J (1991) Microbiological aspects of modified-atmosphere packaging technology – A review. J Food Protect 54, 58–70
Garcia J, Streif J (1993) Qualität und Haltbarkeit von Birnen. I. Einfluß von CA- bzw. ULO-Lagerbedingungen. Gartenbauwiss 58, 36–41
Gill C, Harrison J (1989) The storage life of chilled pork packaged under carbon dioxide. Meat Sci 26, 313–324
Gill C, Molin G (1991) Modified atmospheres and vacuum packaging. In: Russel N, Gould G: Food Preservatives. Blackie, Glasgow, p. 172–199
Gupta S, Subrata B, Panduranga C, Chakraborti R (1991) Preservation of mullets in carbon dioxide enriched gas-mixture at 6–7 °C. Fishery Technol 28, 125–127
Haas G, Prescott H, Dudley E, Dik R, Hintlian C, Keane L (1989) Inactivation of microorganisms by carbon dioxide under pressure. J Food Safety 9, 253–265
Hanninen M, Korreala H, Pakkala P (1984) Effect of various gas atmospheres on the growth and survival of *Campylobacter jejuni* on beef. J Appl Bacteriol 57, 89–94
Henschler D (1992) Toxikologisch-arbeitsmedizinische Begründungen von MAK-Werten, loose-leaf collection. VCH Verlagsgesellschaft, Weinheim
Holley R, Delaquin P, Rodrigue N, Doyon G, Gagnon J, Garipy C (1994) Controlled-atmosphere storage of pork under carbon dioxide. J Food Protect 57, 1088–1093
Kim D, Baek H, Yoon H, Kim K (1989) Effects of carbon dioxide concentration in CA conditions on the quality of Shiitake mushroom (*Lentinus edodes*) during storage. Korean J Food Sci Technol 21, 461–467
Kosikowski F, Brown D (1973) Influence of carbon dioxide and nitrogen on microbial solutions and shelf life of cottage cheese and sour cream. J Dairy Sci 56, 12–18
Krämer K, Baumgart J (1992) Brühwurstaufschnitt. Hemmung von *Listeria monocytogenes* durch modifizierte Atmosphären. Fleischwirtschaft 72, 666, 668, 758
Kubo Y, Inaba A, Nakamura R (1990) Respiration and C_2H_4 production in various harvested crops held in carbon dioxide enriched atmospheres. J Am Soc Horticul Sci 115, 975–978
Lambert A, Smith J, Dodds K (1991) Effect of head space carbon dioxide concentration on toxin production by *Clostridium botulinum* in MAP, irradiated fresh pork. J Food Protect 54, 588–592
Locatelli D, Daolio E (1993) Effectiveness of carbon dioxide under reduced pressure against some insects infesting packaged rice. J Stored Prod Research 29, 81–87
Löwenadler J, Rönner U (1994) Determination of dissolved carbon dioxide by coulometric titration in modified atmosphere systems. Lett Appl Microbiol 18, 285–288
Lubieniecki-von Schelhorn M (1974) Vermehrung und Absterben von Mikroorganismen in Abhängigkeit von Milieu unter besonderer Berücksichtigung kombinierter technologischer Einflüsse. 4. Mitteilung. Chem Mikrobiol Technol Lebensm 3, 138–147
Mazza G, Siemens A (1990) Carbon dioxide concentration in commercial potato storage and its effect on quality of tubers for processing. Am Potato J 67, 121–132
Molin G (1983) The resistance to carbon dioxide of some food related bacteria. J Appl Microbiol Biotechnol 18, 214–217
Penney N, Hagyard C, Bell R (1993) Extension of shelf life of chilled sliced roast beef by carbon dioxide packaging. Int J Food Sci Technol 28, 181–191
Rau G (1993) Alternative Verfahren mit inerten Gasen im Vorratsschutz. Dtsch Lebensm Rundsch 89, 216–219

Shih C, Marth E (1973) Aflatoxin produced by *Aspergillus parasiticus* when incubated in the presence of different gases. J Milk Food Technol 36, 421–425
Smith J, Khanizadeh S, van de Voort F, Hardin R, Ooraikul B, Jackson E (1988) Use of response surface methodology in shelf life extension studies of a bakery product. Food Microbiol 5, 163–176
Troost G (1988) Technologie des Weines. Ulmer, Stuttgart, p. 166
Vora H, Sidhu J (1987) Effects of varying concentrations of ethyl alcohol and carbon dioxide on the shelf life of bread. Chem Mikrobiol Technol Lebensm 11, 56–59
Wei C, Balaban M, Fernando S, Peplow A (1991) Bacterial effects of high pressure carbon dioxide treatment on foods spiked with listeria or salmonella. J Food Protect 54, 189–193J, Bouchard C, Simard R, Pichard B, Holley R (1984) Effect of N_2, CO, CO_2 on the growth of bacteria from meat products under modified atmospheres. Microbiol Aliments Nutri 2, 351–370

Chapter 8

Nitrogen

8.1 Synonyms

IUPAC: Nitrogen
German: Stickstoff, Nitrogenium. *French:* Nitrogène. *Italian:* Azoto. *Spanish:* Azoe, nitrógeno. *Russian:* Азот.

8.2 Commercially Available Forms

Nitrogen is marketed in liquid form or as gas under a pressure of some 150–200 bar.

8.3 Properties

N_2, atomic mass 14.01, at room temperature, a colorless and odorless gas which liquefies under normal pressure at –196 °C. Unlike carbon dioxide, nitrogen is virtually insoluble in water.

8.4 Production

Nitrogen is obtained industrially by fractionation of liquid air.

8.5 Health Aspects

As a largely inert gas, nitrogen is of no toxicological significance apart from the fact that it must not be inhaled in such concentrations that the supply of oxygen falls below the concentration essential to life.

8.6 Regulatory Status

In most countries nitrogen is subject to scarcely any legal restrictions. In the European Union, however, it is defined as an additive when it comes directly in contact with foods, i. e. as a protective gas.

8.7 Antimicrobial Action

The action of nitrogen is based exclusively on its displacement of oxygen essential to obligate aerobic microorganisms. In contrast to carbon dioxide, nitrogen itself has no direct antimicrobial action (Huffmann 1974) (Eklund and Jarmund 1983) and does not even inhibit clostridia (Parekh and Solberg 1970). Aflatoxin formation by molds, however, is hindered by nitrogen although not to the same extent as by carbon dioxide (Shih and Marth 1973). Like carbon dioxide, nitrogen is used as a protective gas. Packages employing nitrogen as a protective gas need to fulfil the same requirements as those in which carbon dioxide is used (see Sect. 7.9.1).

8.8 Fields of Use

Nitrogen is frequently used in mixtures with carbon dioxide, for example as a protective gas for packaged meat and poultry, fish, hard cheese, eggs and baked goods (see Table 13) (Smith et al. 1983, Corinth 1985).

Table 13. Examples of gas mixtures used for various MAP products (Farber 1991)

Product	% CO_2	% O_2	% N_2
Fresh meat	30	30	40
	15–40	60–85	0
Salt meat	20–50	0	50–80
Chilled sliced roast beef	75	10	15
Eggs	20	0	0
	0	0	100
Poultry	25–30	0	70–75
	60–75	5–10	20
	100	0	0
	20–40	60–80	0
Pork	20	80	0
	0	0	100
Fish	40	30	30
	40	0	60
	60	0	40
Hard cheese	0–70	0	30–100
	30	0	70
Sandwiches	20–100	0–10	0–100
	0	0	100
Pastry goods	70–80	0	20–30
Baked goods	20–70	0	80–20
	0	0	100
	100	0	0

8.9 Other Effects

By displacing oxygen, which is usually detrimental to the keeping properties of foods, nitrogen displays what is, in a certain sense, an "antioxidative" effect. Nitrogen is used, for instance, in the packaging of fat products, dairy produce, snacks and other dry products sensitive to oxidation, as well as drinks. The "antioxidative" effect of nitrogen is generally of much greater importance than its indirect antimicrobial action but lies outside the scope of this book.

8.10 Literature

Corinth H-G (1985) Haltbarkeitsverlängerung durch Anwendung von Schutzgas. Lebensmitteltechn 12, 630–684
Eklund T, Jarmund T (1983) Microculture model studies on the effect of various gas atmospheres on microbial growth at different temperatures. J Appl Bacteriol 55, 119–125
Farber J (1991) Microbiological aspects of modified-atmosphere packaging technology, a review. J Food Protect 54, 58–70
Huffmann DL (1974) Effect of gas atmospheres on microbial quality of pork. J Food Sci 39, 723–725
Parekh KG, Solberg M (1970) Comparative growth of *Clostridium perfringens* in carbon dioxide and nitrogen atmospheres. J Food Sci 35, 156–159
Shih CN, Marth EH (1973) Aflatoxin produced by *Aspergillus parasiticus* when incubated in the presence of different gases. J Milk Food Technol 36, 421–425
Smith J, Jackson E, Ooraikul B (1983) Microbiological studies on gas-packaged crumpets. J Food Protect 46, 279–283

Chapter 9

Nitrates

9.1 Synonyms

IUPAC: Potassium nitrate
English: Saltpeter (Am.), saltpetre (Brit.), nitre, niter. *German:* Nitrate, Salpeter. *French:* Nitrates, salpêtre, nitre. Italian: Nitrati, salnitro. *Spanish:* Nitratos, azoatos, salitre, nitro. *Russian:* Нитраты, селитра.

9.2 History

Nitrate has been used for centuries to pickle meat and as an additive for fish and cheese. No one knows for certain who first employed nitrate in food preservation. According to the literature, the Dutchman, Gillis Beukel (d. 1397), was the first to use nitrate for the preservation of fish, and the word "pickle" is said to derive from his name, but this has not been substantiated. Beukel may possibly have worked only with common salt and been unfamiliar with nitrate, like others who had practiced pickling even earlier (Binkerd and Kolari 1975). Nitrate was definitely known by about 1500, as it is mentioned by Sebastian Brandt (as saltpeter) in his satirical poem "The Ship of Fools" (see Sect. 4.1).

Despite more recent revelations concerning the action of nitrites, nitrates remain important to this day, especially in the treatment of meat in large cuts, although there is evidence that its use in food preservation is on the decline.

9.3 Commercially Available Forms

Sodium and potassium nitrate (or sodium and potassium saltpeter, as they may still occasionally be known) are used, either pure or in admixtures with common salt and other substances, as a curing salt.

9.4 Properties

$NaNO_3$, molar mass 84.99, KNO_3, molar mass 101.11, white crystalline powders which melt at 311 °C ($NaNO_3$) and 337 °C (KNO_3). Sodium nitrate is hygroscopic. At room temperature, 100 g water will dissolve about 90 g $NaNO_3$ and about 37 g KNO_3. Both nitrates are only very sparingly soluble in alcohol.

9.5 Analysis

In the detection and determination of nitrates it should be borne in mind that most reactions of the nitrate ion are affected by the presence of nitrite ions, which are frequently found alongside nitrates. This disturbing effect is prevented by removing the nitrite beforehand, e.g. by reacting it with urea in an acid solution or eliminating it with amidosulfonic acid. Besides the conventional color reactions, most of which are the result of reducing the nitrate to nitrite and reacting the nitrite with sulfanilic acid or α-naphthylamine, for example, ion chromatography can be used to identify and quantify nitrates and nitrites in a single run (Stein et al. 1988).

9.6 Production

Sodium nitrate in pure form is obtained by passing the nitrous gases from ammonia combustion through caustic soda or soda solution. Potassium nitrate is produced by reacting potassium carbonate or caustic potash with nitric acid, or from potassium chloride and nitric acid in the presence of oxygen.

9.7 Health Aspects

9.7.1 Acute Toxicity

The LD_{50} of sodium nitrate for rats is in the region of 3–7 g/kg body weight. Nitrates appear to be considerably more toxic to larger animals. The lethal dosage quoted for humans is 30–35 g/kg body weight (Battelle 1972) but, even in substantially smaller doses than this, potassium nitrate causes local intestinal irritation and diarrhea owing to dehydration in the intestine. Potassium nitrate is more toxic than sodium nitrate (Wirth et al. 1985).

9.7.2 Subchronic Toxicity

There has been a lack of systematic feeding experiments relating to the use of nitrates as food additives. Relevant data available is concerned mainly with the intake of nitrates via drinking water or vegetable feed consumed by livestock. Although as little as 100 mg potassium nitrate per liter drinking water causes an increase in the methemoglobin formation in livestock over a 5-week period, other data show that some 2% nitrate, relative to grass feed, has no untoward influence on sheep (Battelle 1972).

In concentrations of up to 400 mg/kg body weight, nitrates have no teratogenic action (Battelle 1972).

9.7.3
Chronic Toxicity

No systematic chronic feeding trials with nitrates are described in the literature (Battelle 1972).

In a carcinogenicity study a NOEL for rats of 2500 mg sodium nitrate/kg body weight per day was obtained. To transfer this finding to humans it is necessary to bear in mind differences between humans and rats in nitrate kinetics (no recirculation of nitrate in the saliva of rats) and the endogenous nitrate formation rate (Walker 1990, Wettig et al. 1990). On the basis of the NOEL of 2500 mg/kg body weight per day and the usual additional safety factor of 100 to 500, the SCF and JECFA have set an ADI of 0–5 mg/kg body weight (SCF 1990). Methemoglobinemia is a particular risk to infants and small children because of their extra sensitivity to nitrate and the nitrite endogenously formed from it. For this reason the maximum permissible quantities for baby food are very low.

According to current knowledge, there is no epidemiological link between the absorption of nitrates and the incidence of cancer. Studies from China, Chile, Japan, and Columbia demonstrating such a connection are encumbered with serious elements of uncertainty and take no account of accompanying vitamin C deficiency, for example (SCF 1990).

9.7.4
Biochemical Behavior

Nitrates may be converted into nitrites by enzymatic processes or by the activity of microorganisms. This reduction is largely uncontrolled – it may occur in foods or in the human digestive tract. In adults, the nitrate is converted to nitrite in the intestine, whereas in infants conversion already occurs in the stomach or duodenum, where nitrite is absorbed especially rapidly. Consequently nitrates are particularly toxic to infants. Although nitrates as such are fairly harmless, the possibility of endogenous nitrite formation and consequent potential for nitrosation must be borne in mind. As a result, the possible formation of nitrosamines cannot be excluded.

9.8
Regulatory Status

Sodium nitrate and in some cases also potassium nitrate are permitted in many countries as additives to certain meat and fish products, as well as to certain types of cheese. Because the extent to which nitrate will be converted to nitrite is not fully predictable, the use of nitrate in meat goods is prohibited in a number of countries, which prefer to allow only the use of nitrite on its own.

9.9 Antimicrobial Action

The antimicrobial action of nitrates is directed exclusively towards anaerobic bacteria. As far as aerobic bacteria are concerned, their growth actually tends to be promoted by nitrates; indeed, a number of microorganism strains can use nitrates as a source of nitrogen. At the concentrations at which nitrates are usually added to food, little in the way of a direct inhibitory action on bacteria can be expected. The antimicrobial action of nitrate derives much more from the nitrite produced from the nitrate in foods.

9.10 Fields of Use

9.10.1 Dairy Produce

Sodium nitrate or potassium nitrate is added in concentrations of 0.01 to 0.02% to vat milk, thus preventing blowing in hard cheese produced from the milk. Higher concentrations are undesired, because they may cause discoloration of the cheese. The nitrate does not act as such in the cheese, where it is more or less rapidly decomposed, depending on the ripening temperature (Schulz et al. 1960). The action is attributable to the nitrite, which occurs as an intermediate product but is no longer detectable in the cheese when this is ready for consumption. This nitrite inhibits the undesired fermentations caused by clostridia, coli or butyric acid bacteria (Schulz et al. 1960). Nitrates are less suitable for combating late blowing in Emmentaler because they may inhibit the development of the desired propionic acid bacteria.

9.10.2 Meat Products

Nitrate in meat products is converted into nitrite by the action of bacteria. This brings about the desired change in the meat's red color, provides the specific "cured" flavor and has an antimicrobial action. The antibacterial action of nitrate itself is only slight (Hustad et al. 1973): in the concentrations employed in practice, only the nitrite formed from it has an antibacterial effect (Leistner et al. 1973). One drawback of employing nitrate is that its conversion into nitrite is uncontrolled.

Nitrate is used dry or in solution. In dry curing, liberal quantities of dry curing salt – a mixture of nitrate, common salt and sucrose – are rubbed into the meat, after which the cuts of meat so treated are placed in pickling vessels and weighted down. In this way the common salt, by means of osmosis, extracts some of the tissue fluid partly covering the meat. Frequently this process is followed by pickle-curing.

Pickle-curing involves placing the meat in pickling brines – solutions of nitrate, common salt and sucrose. The curing temperature is between +6 and +8 °C.

Higher temperatures encourage the reproduction of undesired microorganisms and may thereby cause decomposition in the brines.

9.10.3 Fish Products

Nitrates are added to anchovies not on account of their antimicrobial action but because of their color-retentive properties.

9.11 Literature

Battelle Columbus Laboratories (1972) GRAS (Generally Recognized as Safe) food ingredients – nitrates and nitrites (including nitrosamines). PB 221220. National Technical Information Service, US Department of Commerce, Springfield

Binkerd EF, Kolari OE (1975) The history and use of nitrate and nitrite in the curing of meat. Food Cosmet Toxicol 13, 655–661

Hustad GO, Cerveny JG, Trenk H, Deibel RH, Kautter DA, Fazio T, Johnston RW, Kolari OE (1973) Effect of sodium nitrite and sodium nitrate on botulinal toxin production and nitrosamine formation in wieners. Appl Microbiol 26, 22–26

Leistner L, Hechelmann H, Uchida K (1973) Welche Konsequenzen hätte ein Verbot oder eine Reduzierung des Zusatzes von Nitrat und Nitritpökelsalz zu Fleischerzeugnissen? Aus mikrobiologischer Sicht. Fleischwirtschaft 53, 371–375 and 378 (1990) Bericht des Wissenschaftlichen Lebensmittelausschusses über Nitrate und Nitrite. Stellungnahme vom 19.10.1990, EUR 13913

Schulz ME, Kay H, Mrowetz G (1960) Untersuchungen über das Verhalten von Nitrat im Käse. Milchwissenschaft 15, 556–559

Stein C, Classen H-G, Schwedt G (1988) Kinetic studies on nitrite and nitrate in rats by ion pair chromatography. Clinico Chim Acta 175, 167–175

Walker R (1990) Nitrates, nitrites and N-Nitrosocompounds: a review of the occurrence in food and diet and the toxicological implications. Food Additives Contam 7, 717–768

Wettig K, Broschinski L, Diener W, Fischer G, Flentje B, Namaschk A, Scheibe J, Schulz K-R (1990) Endogen gebildete Nitrate als Teil der gesamten Nitratbelastung des Menschen. Bundesgesundheitsbl 33, 235–240

Wirth W, Gloxhuber C, Krienke EG, Wirth KE (1985) Toxikologie, 4th edn. Georg Thieme, Stuttgart, p. 93

Chapter 10

Nitrites

10.1 Synonyms

German: Nitrite. *French:* Nitrites. *Italian:* Nitriti. *Spanish:* Nitritos. *Russian:* Нитриты.

10.2 History

Nitrates were used in food preservation for centuries (Binkerd and Kolari 1975) before it was eventually realized in 1899 that basically the active agent is not nitrate but the nitrite formed microbiologically from it (Kisskalt 1899). Because the conversion of nitrate to nitrite is a largely uncontrolled process, there has since been increasing preference for the direct use of nitrite. Initially nitrite was used alone but in most countries is now employed only in admixture with common salt in fixed and often legally stipulated ratios. In view of the toxicological properties of the nitrosamines formed under certain conditions by nitrite and those of the nitrite itself, the continued use of nitrites in food preservation is now undergoing critical examination (Walker 1990).

10.3 Commercially Available Forms

The nitrite employed is almost invariably sodium nitrite. In a number of countries sodium nitrite is used only in the form of what is known as nitrite curing salt, i.e. in admixture with common salt. This is done to make dosage simpler and more reliable. The storage stability and accuracy of the blend proportions of such products are generally good (Brauner-Glaesner and Ristow 1984).

10.4 Properties

$NaNO_2$, molar mass 69.00, white to pale yellowish hygroscopic crystals which melt at 306 °C. Sodium nitrite is very readily soluble in water but only sparingly soluble in alcohol (Tiedke and Koch 1992).

10.5 Analysis

For qualitative detection and quantitative determination of nitrite in foods, an aqueous extract of the substance under investigation can be diluted with sulfanilic acid so that it forms a diazonium compound which couples with α-naphthylamine to give a pink azo dye that can be evaluated colorimetrically (Usher and Telling 1975). The reaction with *m*-phenylene diamine can also be used for quantitative nitrite determination (Usher and Telling 1975). Quantitative determination of nitrite by ion chromatography is also possible (Stein et al. 1988, Kim and Conca 1990, Dennis et al. 1990). The nitrite content of nitrite curing salt can be determined by cerimetry or spectrophotometry (Frommberger 1985). Another standardized method of nitrite determination has been published by the AOAC (Helrich 1990).

10.6 Production

Sodium nitrite occurs together with sodium nitrate when the nitrous gases deriving from the combustion of ammonia are passed into caustic soda solution. Nitrite is less readily soluble than nitrate and can be separated out by cooling the solutions.

10.7 Health Aspects

10.7.1 Acute Toxicity

The LD50 of nitrite for rodents is in the region of 100–200 mg/kg body weight (Smyth et al. 1969, Battelle 1972, Burden 1961). For humans, nitrite is rather more toxic, the quoted lethal dosage being 32 mg/kg body weight, corresponding to about 2 g per individual (Burden 1961). According to other data, the lethal dosage is 4 to 6 g (Wagner 1956). The relatively powerful toxicity of pure nitrite has led to poisoning when nitrite has been used improperly as such or mistakenly instead of nitrate in meat technology. Nitrite is therefore today generally employed only in admixture with common salt.

10.7.2 Subchronic Toxicity

When 1.4 g sodium nitrite per liter drinking water was fed to experimental animals over 200 days, this produced an increased methemoglobin content in the blood, as well as changes in the liver, spleen, kidneys and mycocardium (Battelle 1972). Similar changes occur when the animals are fed potassium nitrite. From the results of a feeding study, in which 3000 mg per liter drinking water was fed to rats over a period of 13 weeks, a NOEL of under 100 mg/l was derived. This is equivalent to less than 10 mg/kg body weight per day (Til et al. 1988).

In vitro, nitrite has a mutagenic effect on microorganisms and mammal cells (Walker 1990). Some in-vivo mutagenicity tests produced positive results, others negative ones (SCF 1990).

10.7.3 Chronic Toxicity

The administration of 100 mg nitrite/kg body weight to rats in their feed over three generations lowers the hemoglobin concentration in the blood and reduces the life span to an extent probably not statistically significant. The reproductive capacity of the rats is not impaired by 100 mg nitrite/kg body weight (Druckrey et al. 1963). The administration to guinea-pigs of 5,000 to 10,000 ppm nitrite, relative to the drinking water, over 4 months and corresponding to some 120 and 290 mg nitrite/kg body weight respectively, does not cause any macroscopic changes except to the blood picture, but substantial impairment of the reproductive capacity was observed (Sleight and Atallah 1968). Nitrite is not teratogenic (Druckrey et al. 1963).

The long-term administration of a feed containing 40% meat which had been treated with 0.02 to 0.5% nitrite gave no cause to suspect carcinogenic effects (Logten et al. 1972). Sodium nitrite appears not to be intrinsically carcinogenic, as demonstrated by a feeding experiment in which feed to which up to 0.5% sodium nitrite had been added was administered to rats over 115 weeks (Grant and Butler 1989). A problematic aspect, however, is the possible formation of carcinogenic *N*-Nitroso compounds of nitrite and amines. This can occur both in foods (exogenous nitrosation) and after the intake of nitrite and amines in the stomach (endogenous nitrosation). In addition, further nitrite may be formed by microbial reduction of nitrate. If nitrate is recirculated with the saliva and then swallowed again, this increases the exposure time compared with that of nitrite/nitrate ingested with the feed. Since *N*-nitrosamines are highly carcinogenic the intake of nitrite along with the feed should be kept to a minimum. It is also recommended that substances inhibiting nitrosation should be consumed, e.g. vitamins C and E. Having regard to the considerations just mentioned, the SCF has set a temporary ADI of 0 – 0.1 mg nitrite/kg body weight. This figure does not apply to infants aged less than 3 months. JECFA has set an ADI of 0 – 0.2 mg nitrite/kg body weight and opposes any addition of nitrite to baby food.

10.7.4 Biochemical Behavior

Nitrites are readily absorbed by the intestinal tract. They reduce the tonus of the smooth musculature, thus causing vasodilation and a lowering of the blood pressure, an effect which is being evaluated for therapeutic purposes. In higher dosages nitrites form methemoglobin and thus lead to cyanosis.

However, these effects are irrelevant to the concentrations used in the food sector.

10.8 Regulatory Status

Nitrites (E 249 – E 250) are permitted in a number of countries for curing fish products.

In virtually all countries nitrites are permitted as curing agents and preservatives for meat products. Owing to the toxicity of pure nitrite, a number of countries permit the use of nitrite only in admixture with common salt; in the Federal Republic of Germany only nitrite curing salt may be employed in the manufacture of meat and sausage products. Nitrite curing salt is a mixture of sodium nitrite and common salt which contains at most 0.5% and at least 0.4% sodium nitrite.

10.9 Antimicrobial Action

10.9.1 General Criteria of Action

The antimicrobial action of nitrites is based on the nitrous acid they release and the oxides of nitrogen produced from the nitrous acid. These attach to the amino groups of the dehydrogenase system of the microorganism cell and thus cause an inhibitory action (Quastel and Woolridge 1927). Nitrites also have a specific inhibitory action on bacterial enzymes, which catalyze glucose degradation (Woods et al. 1981, Woods and Wood 1982). There are other points of attack for the nitrite in the bacterial metabolism through which the inhibition of growth can be explained, e.g. reactions with hemoprotein, such as cytochromes and SH enzymes (Castellani and Niven 1955, Roberts et al. 1990).

The action of nitrites increases with a falling pH value, in other words with the acid content of the medium. Whilst 4000 ppm nitrite are required to inhibit *Staphylococcus aureus* at pH 6.9, this minimum inhibitory concentration falls to 400 ppm at pH 5.8 and to 80 ppm at pH 5.05 (Castellani and Niven 1955). An addition of acidulants, glucono delta lactone or inoculation with acid-forming lactobacilli will therefore have a beneficial effect on the action of nitrite.

The nitrite concentrations of 80–160 mg/kg, which are usual in meat technology, are not adequate in the culture medium test to inhibit bacteria reliably. Apparently, an adequate action for practical conditions is obtained only by the combination of nitrite with common salt (Baird-Parker and Baillie 1973), an a_w value reduced by this and other factors, a sufficiently low pH value, a low redox potential and storage temperature, the heating and the low microbe count of foodstuff to be preserved (McLean et al. 1968, Leistner et al. 1973, Ala-Huikku et al. 1977, Lee et al. 1978, Lechowich et al. 1978).

One main point of practical interest in food preservation is the inhibitory action of nitrites on clostridia and thus on the formation of botulinum toxin. This action is increased approximately tenfold as is that against spores of *Clostridium botulinum* (Roberts and Smart 1974) if the nitrite has been heated together with the culture medium (Perigo effect) (Perigo et al. 1967, Perigo and Roberts 1968, Riha and

Solberg 1975a, Riha and Solberg 1975b). The minimum inhibitory concentration of the nitrite is then in the range of 50–200 mg/kg, according to the pH value (Hustad et al. 1973, Baird-Parker and Baillie 1973, Leistner et al. 1973, Grever 1973). The causes of the Perigo effect have still not been completely clarified. Model experiments suggest that reaction products of the nitrite with the contents of meat, which may form upon heating, have an antibacterial action far higher than that of the nitrite itself. Such possible reaction products are nitrosothiols and the reaction products of the nitrite with sulfur-containing compounds and Fe^{2+} of the Roussin's salt type (Mirna and Coretti 1974, Tompkin 1993), although these are rather susceptible to heat (Mirna and Coretti 1974, Moran et al. 1975). Consequently, consideration has now been given to the possibility that other reaction products of nitrite with substances contained in meat might also be responsible for the increased action of the nitrite in heated meat products, e.g. S-nitrosocysteine, complexes of cysteine with Fe^{2+} and oxides of nitrogen (Moran et al. 1975) or reaction products of the nitrite with sugars, amino sugars, sugar aldehydes and other carbonyl compounds (Mirna and Coretti 1976).

10.9.2 Spectrum of Action

The growth of fungi and yeasts is not affected by nitrites, whose action is almost exclusively antibacterial.

10.10 Fields of Use

Nitrite is added to meat products, especially sausages and cured meats, not only to obtain the desired cured color and the specific cured flavor but also to improve the keeping properties against bacterial spoilage. In the context of this book, only the latter effect is of interest. The addition of nitrite to meat products prevents not only the development of pathogenic and toxic microorganisms but also the formation of enterotoxins and other bacteriotoxins. Nitrite thus acts as a preventive against food poisoning. The desired antibacterial action is obtained in practice by concentrations of 50 to 160 mg/kg in the food to be preserved. A reduction in this applied concentration of nitrite, which would be desirable for toxicological reasons, can be justified only if equally multifunctional replacement substances are found. No such substances have yet been discovered.

Of the many products tested, sorbic acid and sorbates are the most suitable (Tóth 1983, Lück 1984). Under the production conditions for pickled goods, i.e. the usual a_w and pH values, the inhibitory effect of sorbic acid and sorbates on pathogenic and toxic microorganisms is in some cases even superior to that of nitrites (see chapter 19). However the pickling color and flavor resulting from the chemical effect of nitrite are not obtained with sorbic acid and sorbates. For this reason sorbates have not so far become established in the sector.

The other main alternatives to nitrite are the Wisconsin process, in which *Pediococcus acidilacticus* is added as a protective culture together with sucrose

(Tanaka et al. 1985), the use of nitrite-free curing systems based on dinitrosylferrohemochrome or protoporphyrin IX (Smith and Burge 1987, O'Boyle et al. 1990, Shahidi and Pegg 1992) or the use of monascus pigments (Fink-Gremmels et al. 1991, Wirth 1991, Leistner 1994). Some of these methods produce an attractive color but in most cases the antimicrobial effect of nitrite is not achieved, or there are technical problems, or the products have not undergone full toxicological testing.

Nitrite is a highly reactive chemical and therefore converted in meat relatively rapidly (Woolford et al. 1976, Woolford and Cassens 1977). Part of it is oxidized to nitrate or converted to oxides of nitrogen, part is linked to myoglobin or other protein, part is bound to amino acids, such as tryptophan or tyrosine (Woolford et al. 1976), and part reacts with SH compounds. Lipids and carbohydrates, too, as well as many other food constituents, are possible reaction partners with nitrite (Greenland 1978).

While dry curing and pickle-curing is carried out mainly with nitrate, nitrite is employed to a greater extent for sausages and other products prepared from comminuted meat. Frequently, nitrite curing is also cheaper than that with nitrate because the curing process is more rapid. Nitrite is also used for rapid curing, including artery pumping, spray pumping, ultrasound curing, and curing under vacuum.

10.11 Other Effects

Nitrite attaches itself to the muscle color myoglobin to form nitrosomyoglobin, which is resistant to boiling. This process provides the red color of pickled meat.

Table 14. Inhibitory action of nitrite on bacteria (Castellani and Niven 1955)

Name of the bacteria	Minimum inhibitory concentration of nitrite in ppm after heating under	
	anaerobic conditions	aerobic conditions
Streptococcus mitis	40	4000
Streptococcus lactis	6000	10000
Streptococcus liquefaciens	800	6000
Streptococcus faecalis	4000	6000
Streptococcus salivarius	80	4000
Streptococcus pyogenes	2	20
Lactobacillus casei	4000	8000
Lactobacillus arabinosus	8000	25000
Pediococcus cerevisiae	8000	25000
Bacillus megatherium	80	4000
Escherichia coli	2000	4000
Aerobacter aerogenes	2000	4000
Proteus vulgaris	400	4000
Salmonella typhosa	800	2000
Salmonella typhimurium	2000	4000
Shigella flexneri	100	2000

In addition, nitrite is involved in producing the desired flavor of cured meat products (Tóth 1983) and protects the fat attached to the meat from oxidative spoilage. These "side effects" are regarded in food technology as being at least as beneficial as the preservative effect.

10.12 Literature

Ala-Huikku K, Nurmi E, Pajulahti H, Raevuori M (1977) Effect of nitrite, storage temperature and time on *Clostridium botulinum* type A. toxin formation in liver sausage. Eur J Appl Microbiol 4, 145-149

Baird-Parker AC, Baillie MAH (1973) The inhibition of *Clostridium botulinum* by nitrite and sodium chloride. Proc Int Symp Nitrite Meat Prod Zeist, p. 77-90

Battelle Columbus Laboratories (1972) GRAS (Generally Recognized as Safe) food ingredients - nitrates and nitrites (including nitrosamines). PB 221220. National Technical Information Service, US Department of Commerce, Springfield

Binkerd EF, Kolari OE (1975) The history and use of nitrate and nitrite in the curing of meat. Food Cosmet Toxicol 13, 655-661

Brauner-Glaesner G, Ristow R (1984) Lagerfähigkeit von Nitritpökelsalz. Lebensmittelchem Gerichtl Chem 38, 91-94

Burden EHWJ (1961) The toxicology of nitrates and nitrites with particular reference to the potability of water supplies. A review. Analyst (London) 86, 429-433

Castellani AG, Niven CF (1955) Factors affecting the bacteriostatic action of sodium nitrite. Appl Microbiol 3, 154-159

Dennis M, Key P, Papworth T, Pointer M, Massey R (1990) The determination of nitrate and nitrite in cured meat by HPLC/UV. Food Add Contam 7, 455-461

Druckrey H, Steinhoff D, Beuthner H, Schneider H, Klärner P, (1963) Prüfung von Nitrit auf chronisch toxische Wirkung an Ratten. Arzneim Forsch 13, 320-323

Fink-Gremmels J, Dresel J, Leistner L (1991) Einsatz von Monascus-Extrakten als Nitrit-Alternative bei Fleischerzeugnissen. Fleischwirtsch 71, 329-331

Frommberger R (1985) UV-spektrophotometrische Schnellbestimmung des Natriumnitritgehaltes von Nitritpökelsalz. Lebensmittelchem gerichtl Chem 39, 99-101

Grant D, Butler W (1989) Chronic toxicity of sodium nitrite in the male F 344 rat. Food Chem Toxicol 27, 565-571

Greenland S (1978) The interaction of nitrites with food, drugs and contaminants. J Environ Health 41, 141-143

Grever ABG (1973) Minimum nitrite concentrations for inhibition of *Clostridia* in cooked meat products. Proc Int Symp Nitrite Meat Prod Zeist, p. 103-109

Helrich K (ed) (1990) Official methods of analysis of the Association of Official Analytical Chemists. AOAC, Arlington

Hustad GO, Cerveny JG, Trenk H, Deibel RH, Kautter DA, Fazio T, Johnston RW, Kolari OE (1973) Effect of sodium nitrite and sodium nitrate on botulinal toxin production and nitrosamine formation in wieners. Appl Microbiol 26, 22-26

Kim H, Conca K (1990) Determination of nitrite in cured meats by ion-exchange chromatography with electrochemical detection. J Assoc Off Anal Chem 73, 561-564

Kisskalt K (1899) Beiträge zur Kenntnis der Ursachen des Rothwerdens des Fleisches beim Kochen, nebst einigen Versuchen über die Wirkung der schwefligen Säure auf die Fleischfarbe, I. Ueber das Rothwerden des Fleisches beim Kochen. Arch Hyg 35, 11-18

Lechowich RV, Brown WL, Deibel RH, Somers II (1978) The role of nitrite in the production of canned cured meat products. Food Technol 32, 45, 48, 50, 52, 56, 58

Lee SH, Cassens RG, Sugiyama H (1978) Factors affecting inhibition of *Clostridium botulinum* in cured meat. J Food Sci 43, 1371-1374

Leistner L (1994) Die ernährungsphysiologische Bedeutung von Angkak. Fleischwirtsch 74, 772, 775–778
Leistner L, Hechelmann H, Uchida K (1973) Welche Konsequenzen hätte ein Verbot oder eine Reduzierung des Zusatzes von Nitrat und Nitritpökelsalz zu Fleischerzeugnissen? Aus mikrobiologischer Sicht. Fleischwirtschaft 53, 371–375, 378
Logten MJ van, Tonkelaar EM den, Kroes R, Berkvens JM, Esch GJ van (1972) Longterm experiment with canned meat treated with sodium nitrite and glucono-lactone in rats. Food Cosmet Toxicol 10, 475–488
Lück E (1984) Sorbinsäure und Sorbate. Konservierungsstoffe für Fleisch und Fleischwaren. Literaturübersicht. Fleischwirtschaft 64, 727–733
McLean RA, Lilly HD, Alford JA (1968) Effects of meat-curing salts and temperature on production of staphylococcal enterotoxin B. J Bacteriol 95, 1207–1211
Mirna A, Coretti K (1974) Über den Verbleib von Nitrit in Fleischwaren. II. Untersuchungen über chemische und bakteriostatische Eigenschaften verschiedener Reaktionsprodukte des Nitrits. Fleischwirtschaft 54, 507–510
Mirna A, Coretti K (1976) Inhibitory effect of nitrite reaction products and of degradation products of food additives. Proc 2nd Int Symp Nitrite Meat Prod Zeist, p. 39–45
Moran DM, Tannenbaum SR, Archer MC (1975) Inhibitor of *Clostridium perfringens* formed by heating sodium nitrite in a chemically defined medium. Appl Microbiol 30, 838–843
O'Boyle AR, Rubin LJ, Diosady LL, Aladin-Kassam N, Comer F, Brightwell W (1990) A nitrite-free curing system and its application to the production of wieners. Food Technol 5, 88–104
Perigo JA, Whiting E, Bashford TE (1967) Observations on the inhibition of vegetative cells of *Clostridium sporogenes* by nitrite which has been autoclaved in a laboratory medium, discussed in the context of sub-lethally processed cured meats. J Food Technol 2, 377–397
Perigo JA, Roberts TA (1968) Inhibition of *Clostridia* by nitrite. J Food Technol 3, 91–94
Quastel JH, Woolridge WR (1927) The effects of chemical and physical changes in environment on resting bacteria. Biochem J 21, 148–168
Riha WE, Solberg M (1975 a) *Clostridium perfringens* inhibition by sodium nitrite as a function of pH, inoculum size and heat. J Food Sci 40, 439–442
Riha WE, Solberg M (1975 b) *Clostridium perfringens* growth in a nitrite containing defined medium sterilized by heat or filtration. J Food Sci 40, 443–445
Roberts TA, Smart JL (1974) Inhibition of spores of *Clostridium* spp. by sodium nitrite. J Appl Bacteriol 37, 261–264
Roberts T, Woods L, Payne M, Cammack R (1990) Nitrite. In: Russell J, Gould G (eds) Food Preservatives. Blackie, Glasgow p. 89–110
SCF (1990) Stellungnahme vom 19.10.1992 zur Unbedenklichkeit von Nitraten und Nitriten als Lebensmittelzusatzstoffe. 26th report, EUR 13913
Shahidi F, Pegg RB (1992) Nitrite-free meat curing systems: Update and review. Food Chem 43, 185–191
Sleight SD, Atallah OA (1968) Reproduction in the guinea pig as effect by chronic administration of potassium nitrate and potassium nitrite. Toxicol Appl Pharmacol 12, 179–185
Smith JS, Burge DL (1987) Protoporphyrin-IX as a substitute for nitrite in cured-meat color production. J Food Sci 52, 1728–1729
Smyth HF, Carpenter CP, Weil CS, Pozzani UC, Striegel JA, Nycum JS (1969) Range-finding toxicity data: List VII. Am Ind Hyg Assoc J 30, 470–476
Stein C, Classen H-G, Schwedt G (1988) Kinetic studies on nitrite and nitrate in rats by ion-pair chromatography. Clin Chim Acta 175, 167–174
Tanaka N, Meske L, Doyle MP, Traisman E, Thayer DW, Johnston RW (1985) Plant trials of bacon made with lactic acid bacteria, sucrose and lowered sodium nitrite. J Food Protect 48, 679–686
Tiedtke J, Koch H (1992) Nitrit-Pökelsalz-Entwicklungen, lebensmittelrechtliche und produktionsspezifische stoffliche Eigenschaften und Wirkungen. Lebensmittelkontrolleur 7: 10, 18
Til H, Falke H, Kuper C, Willems M (1988) Evaluation of the oral toxicity of potassium nitrite in a 13-week drinking-water study in rats. Food Chem Toxicol 26, 851–859

Tompkin RB (1993) Nitrite. In: Davidson PM, Branen AL: Antimicrobials in Foods, 2nd edn. Marcel Dekker, New York, p. 191–262

Tóth L (1983) Nitrite reactions during the curing of meat products. Fleischwirtschaft 63, 208–211

Usher CD, Telling GM (1975) Analysis of nitrate and nitrite in foodstuffs. A critical review. J Sci Food Agric 26, 1793–1805

Wagner H-J (1956) Vergiftung mit Pökelsalz. Arch Toxikol 16, 100–104

Walker R (1990) Nitrates, nitrites and N-nitrosocompounds: a review of the occurrence in food and diet and the toxicological implications. Food Add Contam 7, 717–768

Wirth F (1991) Einschränkung und Verzicht auf Nitrit bei Pökelstoffen in Fleischerzeugnissen. Fleischwirtsch 71, 228–239

Woods LFJ, Wood JM, Gibbs PA (1981) The involvement of nitric oxide in the inhibition of the phosphoroclastic system of *Clostridium sporogenes* by sodium nitrite. J Gen Microbiol 125, 399–406

Woods LFJ, Wood JM (1982) A note on the effect on nitrite inhibition on the metabolism of *Clostridium botulinum*. J Appl Bacteriol 52, 109–110

Woolford G, Cassens RG, Greaser ML, Sebranek JG (1976) The fate of nitrite: Reaction with protein. J Food Sci 41, 585–588

Woolford G, Cassens RG (1977) The fate of sodium nitrite in bacon. J Food Sci 42, 586–589, 596

Chapter 11

Ozone

11.1 Synonyms

German: Ozon. *French:* Ozone. *Italian:* Ozono. *Spanish:* Ozono. *Russian:* Озон.

11.2 History

Ozone was the first oxidizing agent to be used for disinfecting drinking water, and its original use can be traced back to the 1880s. The drinking water of Nice has been ozone-treated since 1906 (Rice et al. 1981).

Ozone is becoming increasingly important for drinking water treatment because of the growing misgivings concerning drinking water chlorination.

11.3 Properties

O_3, molar mass 48.0, blue gas whose characteristic odor remains perceptible even in extreme dilution ratios.

11.4 Analysis

With o-tolidine, ozone produces a yellow coloration which can be photometrically evaluated and employed for quantitative determination of ozone in water. UV-spectroscopic, colorimetric and chemiluminescent methods can also been used. Rapid and semi-quantitative determination of ozone is possible with special test tubes (Drägerwerk 1991).

11.5 Production

Being largely unstable, ozone is usually manufactured in situ by silent electric discharge from molecular oxygen in so-called ozonizers. Should pure oxygen be used in preference to air for ozone production, double the quantity of ozone can be obtained under otherwise identical conditions (Aquodrei process).

11.6 Health Aspects

Ozone is a highly toxic gas. Even concentrations of as little as 1–2 mg/m^3 in respiratory air can cause irritation to the mucous membranes (Dungworth et al. 1975, Gilgen and Wanner 1966). After exposure for a certain number of hours depending on the animal species, the acute toxicity is between 2 and 25 ppm, relative to the respired air (Mittler et al. 1959, Gilgen and Wanner 1966). The maximum allowable workplace concentration for ozone is in Germany 0.2 mg/m^3. In 1995, ozone was added to the german MAK Committee's list III A2 (definitely carcinogenic in animal tests). Chronic exposure results in increased lipid peroxidation and thus oxidative damage to cellular macromolecules (Sayato et al. 1993). Because ozone is dangerous to man in bactericidally effective concentrations, it is unsuitable as a deodorant or disinfectant in rooms used for human occupation for any length of time.

11.7 Regulatory Status

Ozone is permitted in various countries for the treatment of drinking water, mainly in maximum concentrations up to 10 mg per liter of water. After treatment the maximum quantity permitted is 0.05 mg ozone per liter drinking water.

11.8 Antimicrobial Action

Ozone will kill microorganisms relatively quickly, if used in the required concentrations. It is thus a disinfectant rather than a preservative (Fetner and Ingols 1959, Chaigneau 1977). The antimicrobial action of ozone is based essentially on its powerful oxidizing effect, which causes irreversible damage to the fatty acids in the cell membrane and to cellular macromolecules, such as proteins and DNA (Fetner and Ingols 1959, Hoffman 1971, Naitoh 1994). This action is particularly effective at high relative air humidity, the bacteria apparently being killed by ozone more readily in a swollen state than when dry. Gram-positive bacteria are basically more sensitive to ozone than are gram-negative bacteria. The bactericidal action of ozone is distinctly more powerful than that of free chlorine (Fetner and Ingols 1959). Owing to its strong oxidizing action, ozone may destroy other preservatives (Brigance and Buescher 1993).

11.9 Fields of Use

11.9.1 Drinks

The main field of use for ozone is the disinfection of drinking water. The advantage of using ozone is that it changes into molecular oxygen shortly after killing the microbes; so no undesired foreign substances remain in the water as happens when chlorine is used, for instance. The drinking water to be treated is brought into contact with an ozone/air mixture produced immediately beforehand in ozonizers. The necessary applied concentration is 1 to 5 mg ozone per liter water, according to the microbe count and other factors (Rice et al. 1981, Kurzmann 1983, Kußmaul and Gerz 1993, Ostruschka 1993).

11.9.2
Other Points

Ozone is also used for deodorizing and disinfecting air in cold stores and commercial freezers for vegetables, fruit and meat. The applied concentration is 2 to 3 mg ozone per m^3 air (Kaess and Weidemann 1968 a). In contact with meat, ozone may cause color changes (Kaess and Weidemann 1968 a and b). For salmonella decontamination, poultry in the USA are often sprayed with ozone-containing water. Ozone is also used for sterilizing equipment and bottles in wineries (Date 1994).

11.10
General Literature

N.N. (1959) Ozone Chemistry and Technology. Adv Chem Series No 21 Am Chem Soc.

11.11
Specialized Literature

Brigance A, Buescher R (1993) Effect of ozone on softening enzymes, sorbate, pigment and bacteria in recycled pickle brine. J Food Biochem 16, 359–369
Chaigneau M (1977) Stérilisation et désinfection per les gaz. Sainte-Ruffine: Maisonneuve, p. 195–207
Date S (1994) Ozone sterilisation – why is it used in many European and South African wineries? Austral Grapegrower Winemaker 371, 39 and 41
Drägerwerk (1991) Dräger-Röhrchen Handbuch, Drägerwerk AG, Lübeck
Dungworth DL, Cross CE, Gillespie JR, Plopper CG (1975) The effects of ozone on animals. In: Murphy JS, Orr JR: Ozone chemistry and Technology. A Review of the Literature: 1961–1974. Franklin Institute Press, Philadelphia, p. 27–54
Fetner RH, Ingols RS (1959) Bactericidal activity of ozone and chlorine against *Escherichia coli* at 1 °C. In: Ozone Chemistry and Technology, Adv Chem Series No 21. Am Chem Soc, Washington, p. 370–374
Gilgen A, Wanner HU (1966) Die toxikologische und hygienische Bedeutung des Ozons. Arch Hygiene 150, 62–78
Hoffman RK (1971) Ozone. In: Hugo WB Inhibition and Destruction of the Microbial Cell, Academic Press, London, p. 251–253
Kaess G, Weidemann JF (1968) Ozone treatment of chilled beef. I. Effect of low concentrations of ozone on microbial spoilage and surface colour of beef. J Food Technol 3, 325–334
Kaess G, Weidemann JF (1968) Ozone treatment of chilled beef. II. Interaction between ozone and muscle. J Food Technol 3, 335–343
Kurzmann GE (1993) Ozonanwendung in der Trinkwasseraufbereitung. Expert-Verlag, Ehningen:
Kußmaul H, Gerz R (1993) Ozon in der Wasseraufbereitung. Zweck und Nebenreaktionen. Getränke Ind 9, 656–658

Mittler S, King M, Burkhardt B (1959) Toxicity of ozone. In: Ozone Chemistry and Technology, Adv Chem Series No 21. Am Chem Soc, Washington, p. 344–351
Naitoh S (1994) Inhibition of food spoilage fungi by application of ozone. Japan J Food Microbiol 11, 11–17
Ostruschka M (1993) Die Abtötung von Mikroorganismen in Wasser durch Ozon unter Fließgleichgewichtsbedingungen. Diss Univ Tübingen
Rice RG, Robson CM, Miller GW, Hill AG (1981) Uses of ozone in drinking water treatment. J Am Water Works Assoc 73, 44–57
Sayato Y, Nakamuro K, Ueno H (1993) Toxicological evaluation of products formed by ozonation of aqueous organics. Japan J Tox Environm Health 39, 251–265

Sulfur Dioxide

12.1 Synonyms

German: Schwefeldioxid, Schwefligsäureanhydrid, "schweflige Säure". *French:* Bioxyde de soufre, anhydride sulfureux, gaz sulfureux, "acide sulfureux". *Italian:* Biossido di zolfo, anidride solforosa, "acido solforoso". *Spanish:* Dióxido de sulfuro, anhidrido sulfuroso, "acido sulfuroso". *Russian:* Двуокись серы, серный ангидрид, "сернистая кислота".

12.2 History

Even in some ancient civilizations (Assyria, China and Greece), sulfur dioxide was used as a fumigating agent "for driving out evil spirits", and probably also for disinfection purposes (Homer, Odyssey XXII, 481). The ancient Romans knew "vapor of sulfur" as an improving agent for wine (Plinius, Naturalis historia XIV, 129). In the opinion of other authors (von Bassermann-Jordan 1923), however, the form in which sulfur was employed is a matter of conjecture.

The use of sulfur dioxide most likely became common practice only in the late Middle Ages. At all events, its use clearly gave rise to abuse at an early stage. In Cologne (Germany) the use of sulfur dioxide in wine was banned completely in the fifteenth century on the grounds that "die natur des menschen belästigt und der trinker in krankheit gebracht werde", (i.e., "it abuseth man's nature and afflicteth the drinker") (von Bassermann-Jordan 1923). In 1487 a decree was issued in Rothenburg ob der Tauber (Germany) that, although the addition of sulfur to the barrels would remain permissible, "auf ein füdriges Fass nicht mehr als ein Lot reinen Schwefels zu nehmen sei" (i.e., "no more than half-an-ounce of pure sulfur might be added to each tun"). Moreover, only a single addition of sulfur was permitted to a particular wine. The excessive addition of sulfur to wine was disapproved of in 1497 at the Diet of Lindau and a year later at the Diet of Freiburg im Breisgau (Strahlmann 1974).

In the centuries that followed, sulfur dioxide remained a widely used preservative for a large number of foodstuffs. Although one of the oldest preservatives still in use, sulfur dioxide remains essential for the production of many foods despite several toxicological reservations.

12.3 Commercially Available Forms, Derivatives

Sulfur dioxide is employed as such in liquid form supplied under pressure in cylinders and as aqueous solutions. In addition, various sulfites are of considerable importance.

12.4 Properties

SO_2, molar mass 64.06, at room temperature and under normal pressure a colorless, non-combustible gas with a pungent odor, which readily liquefies to yield a liquid that boils at –10 °C. The density of SO_2 in gaseous form is at least twice that of air. Water will dissolve 80 l/l at 0 °C and 40 l/l at 20 °C.

$Na_2SO_3 \cdot 7H_2O$, molar mass 252.15, K_2SO_3, molar mass 158.27, $NaHSO_3$, molar mass 104.06, $KHSO_3$, molar mass 120.16, $Na_2S_2O_5$, molar mass 190.10, $K_2S_2O_5$, molar mass 222.34, $CaSO_3 \cdot 2H_2O$, molar mass 156.17, white powders which, apart from calcium sulfite, readily dissolve in water and have a more or less powerful odor of sulfur dioxide. Bisulfites ($NaHSO_3$ and $KHSO_3$) exist only in solution form; on drying, these form pyrosulfites ($Na_2S_2O_5$ and $K_2S_2O_5$).

Table 15. Chemical formulas and SO_2 contents of the most important sulfites

Chemical	Formula	Content of active SO_2
Sulfur dioxide	SO_2	100%
Sodium sulfite, anhydrous	Na_2SO_3	50.82%
Sodium sulfite, heptahydrate	$Na_2SO_3 \cdot 7\,H_2O$	25.41%
Sodium hydrogen sulfite	$NaHSO_3$	61.56%
Sodium metabisulfite	$Na_2S_2O_5$	67.39%
Potassium metabisulfite	$K_2S_2O_5$	57.68%
Calcium sulfite	$CaSO_3 \cdot 2\,H_2O$	64.00%

12.5 Analysis

Sulfur dioxide reduces iodate to free iodine. Potassium iodate starch paper turns blue in the presence of SO_2. Iodine starch paper discolors in the presence of SO_2 because this reduces free iodine to iodide.

For quantitative determination of sulfur dioxide in foods the foods in question can be titrated directly with iodine solution unless substances likely to interfere with the results are present. One variant is to add iodide to the food and titrate with iodate solution. A more exact method, although also more time-consuming, is to separate the sulfur dioxide from the material being investigated by boiling with dilute hydrochloric acid and introducing carbon dioxide. The sulfur dioxide that passes over is collected in a solution of hydrogen peroxide and oxidized there to sulfate, which can be determined by alkalimetry, complexometry or weight analysis (Monier-Williams 1927, Reith and Willems 1958). This type of method measures the total sulfur dioxide contained in a food. More recent methods for quantitative determination of sulfur dioxide are flow injection analysis, in which the criterion used for detection is decoloration of malachite green (Sullivan et al. 1990), as well as modified versions of this method (Bendtsen and Jørgensen 1994). Ion chromatography can be used to distinguish between free and bound sulfur dioxide (Warner et al. 1990, Paíno-Campa et al. 1991). The favored detection system for this employs an electrochemical detector (reductive amperometric detection) (Cardwell 1993). Sulfites can also be determined enzymatically with the aid of sulfite oxidase (Cabré et al. 1990). It is possible to determine free and, after appropriate isolation, bound sulfur dioxide by using immobilized *Thiobacillus thiooxidans* as a biosensor (Kawamura et al. 1994). According to a comparative survey of analytical methods for sulfite in current use in 1990 (FIA, colorimetry, enzymatic determination), the Monier-Williams method (Monier-Williams 1927) continues to be the method of choice (Fazio and Warner 1990).

12.6 Production

Sulfur dioxide is produced by heating (roasting) sulfidic ores or, in purer form, by burning elemental sulfur. As a method of purification, either the crude sulfur dioxide is deep-frozen, whereupon pure sulfur dioxide separates out as a liquid, or else the crude sulfur dioxide is washed out with cold water then desorbed again from the solution by heating.

Sulfites are produced by the reaction of sulfur dioxide with the relevant alkaline solutions. Depending on the stoichiometric conditions, solutions of sulfites or bisulfites are formed. The solid sulfites or pyrosulfites are produced from these by evaporation.

12.7 Health Aspects

12.7.1 Acute Toxicity

When sulfur dioxide is administered perorally, the LD_{50} for the rat is 1,000 – 2,000 mg SO_2 per kg body weight. The lower value was obtained with doses of 6.5% solutions and the higher value with doses of 3.5% aqueous solutions (Jaulmes 1970). The acute toxicity of sodium metabisulfite (Lanteaume et al. 1969) is of the same order. For the rabbit the LD_{50} per os was determined as 600 – 700 mg SO_2/kg body weight, and for the cat as 450 mg. In the case of dogs and also humans, fatal poisoning with sulfur dioxide per os is impossible because vomiting occurs (Lang 1960). Sulfur dioxide and sulfites are considerably more toxic when administered intravenously (Hoppe and Goble 1951).

Human reactions to sulfur dioxide vary widely. Whereas a number of persons can tolerate up to 4 g sulfite daily without untoward effects (i.e., some 50 mg/kg body weight), others complain of headaches, nausea, diarrhea or a feeling of satiation after ingesting only very small quantities (Wucherpfennig 1978). Another factor that appears to be of major importance to the tolerability of sulfur dioxide in wine is the condition of the gastric juices; anacid and subacid persons are considerably more sensitive than those with normal acidity of the gastric juice (Schanderl 1956). In principle, bound sulfur dioxide acts in the organism in the same manner as free sulfur dioxide, the differences being merely in the intensity and speed with which the reaction commences and therefore due to varying kinetics (Lang 1960).

12.7.2 Subchronic Toxicity

Administration of 0.5 to 1% potassium metabisulfite to rats over a period of 10 days leads to increased excretion of calcium (Hugot et al. 1965). The toxicity of feed having a content of 0.6% sodium metabisulfite passes through two phases. In the first two months the salient features are vitamin B_1 deficiency symptoms and a slight antithiamine action. The later form of the toxic effect of such feed with 3 to 4 months' storage time can be only partially corrected by vitamin B_1. In addition, diarrhea occurs (Bhagat and Lockett 1964). After three months' absorption of 160 mg sodium bisulfite per kg body weight per day, the mortality rate of mice is increased appreciably, especially under conditions of hunger (Shtenberg and Ignat'ev 1970). Within 10 to 56 days a dose of 6 to 8% sodium metabisulfite in the feed of rats produces substantial growth retardation, as well as a reduction in feed absorption and feed utilization. Anemia occurs above 2%. Even doses of 1% sulfite cause damage to various organs (Rehm and Wittmann 1963).

In a feeding trial conducted on pigs and lasting 15 weeks, only sulfite concentrations of at least 1.7% have an unfavorable effect on the growth and condition of the internal organs. The thiamine concentration in the urine and liver, however, is reduced even with dosages of as little as 0.16% (Til et al. 1972 b).

12.7.3 Chronic Toxicity

Within a year, 0.5–2% sodium bisulfite in the feed of rats produces damage of various kinds in the nervous system, reproductive organs, bone tissue, kidneys and other internal organs. Doses of less than 0.25% produce no pathological symptoms apart from diarrhea with levels of more than 0.1%. In male rats, additions to the feed of less than 0.1% sodium bisulfite actually produce an increase in weight (Fitzhugh et al. 1946). The addition of 0.12% sodium pyrosulfite to the drinking water, which corresponds to an ingestion of 30–90 mg SO_2/kg body weight, was tolerated by rats over 20 months without serious damage (Cluzan et al. 1965). The only abnormalities were an increase in the leukocytes among the male animals, an increase in the spleen weight of the females and a reduction in the number of young per litter. The administration of 350 to 700 ppm sulfur dioxide in the form of sodium metabisulfite over three generations of rats produced no untoward effects on the internal organs, growth, reproductive capacity or weight of the young (Lockett and Natoff 1960). Doses of up to 0.35% sodium metabisulfite fed to pigs over a period of 48 weeks were tolerated without reaction, although doses above 0.83% produce organic damage of various kinds (Til et al. 1972 b).

Experiments on Wistar rats with wine containing 100 mg and, in other cases, 450 mg SO_2/l over four generations revealed no abnormalities with regard to the protein utilization of the feed, reproductive capacity, macroscopic and histological conditions, biochemical behavior or weight of various internal organs. Differences relative to the control animals were revealed only among those rats which had received the wine with the higher dose of SO_2; these animals displayed retarded growth (Lanteaume et al. 1965).

Feed enriched with thiamine and containing 2% sodium metabisulfite retards growth in long-term tests on rats. This also affects subsequent generations. Even a sulfite concentration of as little as 0.5% has a slight influence on the reproductive capacity. Other organic damage was observed only at higher sulfite dosages; below 0.25% no differences whatever were observed in comparison with the control rats (Til et al. 1972 a).

The IARC (International Agency for Research on Cancer) states that there is "inadequate evidence" of sulfites having a carcinogenic effect on humans; there is "limited evidence" of sulfur dioxide having such an effect on experimental animals and "inadequate evidence" in the case of sulfites, bisulfites and metabisulfites (IARC 1992).

Sulfur dioxide may have a mutagenic effect on microorganisms (Mukai et al. 1970, Hayatsu and Miura 1970).

On the basis of this data and taking special account of sulfite-induced intolerant reactions, both the SCF and JECFA have allocated a group ADI for sulfites, calculated as sulfur dioxide, of 0 to 0.7 mg/kg body weight.

12.7.4 Intolerance Reactions

Sulfite can induce genuine allergies and pseudoallergic reactions in humans (Simon and Stevenson 1988). Most intolerance reactions to sulfites take the form of asthmatic attacks or urticaria (Vena et al. 1994). They are frequently accompanied by an intolerance reaction to acetyl salicylic acid. Depending on the sensitivity of the subject, they may be induced by 2 to 250 mg sulfur dioxide (Belchi-Hernandez et al. 1993). For this reason foods containing sulfites should be labeled accordingly (Nagy et al. 1995).

12.7.5 Biochemical Behavior

The oxidation stage of the hexavalent sulfur is the most stable. The quadrivalent positively charged sulfur present in sulfur dioxide and sulfites therefore tends in principle to be converted to sulfate by oxidation. In the organism oxidation from sulfite to sulfate is catalyzed by sulfite oxidase (enzyme list 1.8.3.1), an enzyme that occurs predominantly in the heart, liver and kidneys (Bhagat and Lockett 1960, Gunnison 1981). In addition, there appear to be other, less specific enzymes that also oxidize sulfite to sulfate, e.g. xanthine oxidase. Sulfate is rapidly excreted in the urine. For this reason sulfur dioxide is not accumulated in the body.

Other factors of toxicological importance are the intervention of sulfur dioxide and sulfites in bodily functions as well as their reaction with vitamins and vital enzymes (Gunnison 1981). Thus, SO_2, even in small concentrations, inhibits dehydrogenases (Pfleiderer et al. 1956). Compounds with disulfide bridges, such as cystine, are reduced by sulfites to the corresponding sulfhydril compounds (Lang 1970). In addition, sulfite destroys thiamine by breaking open the bond between the pyrimidine and the thiazole portion of the molecule. Rats whose feed is additionally enriched with thiamine will tolerate considerably higher quantities of SO_2 than control animals receiving no supplementary thiamine (Hötzel et al. 1966). Cytosine and 5-methylcytosine are de-aminated by sulfite in vitro.

12.8 Regulatory Status

Sulfur dioxide, some sulfites, bisulfites, and pyrosulfites are permitted in virtually all countries as food preservatives, especially for the treatment of plant products and wine.

The maximum permissible quantities vary according to the type of food. In the case of foods for direct consumption these seldom exceed 100 mg/kg. The maximum quantities for wine are in the region of 200 to 250 ml/l, depending on the country and type of wine; for some types of wine even greater quantities may be employed.

12.9 Antimicrobial Action

12.9.1 General Criteria of Action

The antimicrobial action of sulfur dioxide is based essentially on the inhibition of enzyme-catalyzed reactions. Its powerful inhibitory effect on enzymes with SH groups has long been known. The great sensitivity of these enzymes to sulfite exists in the primary inhibitory effect of NAD-dependent reactions (Pfleiderer et al. 1956). In the case of yeasts the blocking of the reaction stage of glyceraldehyde 3-phosphate to 1,3-di-P-glycerate is the salient feature, while with *Escherichia coli* it is mainly the NAD-dependent formation of oxalacetate from malate which is inhibited (Wallnöfer and Rehm 1965). In addition, sulfur dioxide inhibits enzymatic reaction cascades by capturing end products or intermediate products as a result of its chemical reactivity. Thus, the acetaldehyde formed in the decomposition of carbohydrate is bound immediately after it is formed. The reaction equilibrium shifts because the addition compounds formed can no longer be attacked by the enzymes (Rehm and Wittmann 1962).

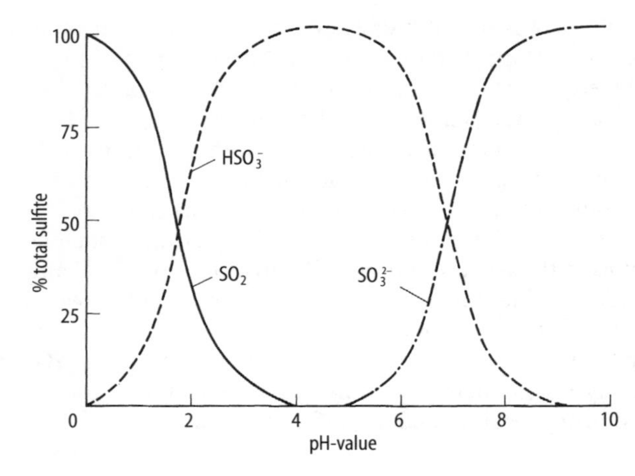

Fig. 8. Proportions of SO_2, HSO_3^- and SO_3^{2-} in aqueous solution as functions of the pH value

With sulfurous acid, as with other preservative acids, the pH value of the product to be preserved is an important factor governing the antimicrobial effect. The position of sulfur dioxide here is a special one, however. Besides dissolved SO_2 gas, three dissociation stages stand together in equilibrium -undissociated sulfurous acid H_2SO_3, hydrogen sulfite HSO_3^- and sulfite ions SO_3^{2-}. Up to a pH value of about 1.7 the content of undissociated sulfurous acid is predominant. In the pH range from 1.7 to 5.1 the main quantity comprises hydrogen sulfite ions, while above pH 5.1 the majority of the sulfurous acid is dissociated. The proportions of the various dissociation stages are shown in Fig. 8 (King et al. 1981).

It is hard to determine in which proportion the sulfurous acid has an antimicrobial action in any particular case (Rehm and Wittmann 1962 and 1963). The most powerful action is that of the dissolved sulfur dioxide gas and undissociated sulfurous acid (Rehm and Wittmann 1963). However, the hydrogen sulfite ions also have an antimicrobial action, even though it is less powerful. This explains the effect of sulfites in the medium and higher pH ranges. The differences between the actions of undissociated sulfurous acid and the hydrogen sulfite ions vary according to the microorganism. The completely dissociated sulfite stage is virtually ineffective, and in this respect the sulfurous acid behaves like other preservative acids.

Sulfurous acid may form addition compounds with food ingredients. Of these, the most important are carbonyl compounds such as aldehydes, ketones and sugars, with which sulfurous acid reacts to form sulfonates. The pH range in which sulfurous acid is used most, namely pH 3–5, is highly conducive to the formation of such compounds (Rehm and Wittmann 1962 and 1963). As a general rule, the linkage of sulfurous acid to carbonyl compounds in foods reduces or even completely neutralizes the antimicrobial action of sulfurous acid against yeasts. In this the acetaldehyde sulfonate is still slightly active, although the addition compounds with sugars are practically ineffectual (Rehm et al. 1964 and 1965). However, the sulfurous acid linked to the acetaldehyde is highly effective against lactobacilli (Fornachon 1963).

12.9.2
Spectrum of Action

Sulfurous acid and its salts belong to the preservatives with a relatively powerful antibacterial action. Its action against bacteria is more pronounced than that against yeasts and molds.

The minimum inhibitory concentrations recorded in Tables 16 to 18 can be transferred to practical conditions only with great reservations since they are based on an observation time of only a matter of hours and no account is taken of the many opportunities of sulfur dioxide to react with food constituents. Lactobacilli are highly sensitive to sulfur dioxide, a fact of great importance for food preservation (Carr et al. 1976).

It will be seen from Tables 16–18 that the inhibitory action of sodium sulfite on yeasts is appreciably lower than that on bacteria. It is also remarkable that many types of yeast react very differently to sulfur dioxide, depending on the strain (Rehm and Wittmann 1962). In similar concentrations, sulfur dioxide is as effective against molds as it is against yeasts (see Table 18).

Table 16. Inhibitory action of sulfur dioxide against bacteria at pH 6 (Rehm et al. 1961, Rehm and Wittmann 1962)

Name of the test organism	Minimum inhibitory concentration of sodium sulfite in ppm
Pseudomonas fluorescens	500
Pseudomonas effusa	500
Pseudomonas ovalis	1000
Staphylococcus aureus	800
Lactobacillus casei	1000
Lactobacillus arabinosus	550
Escherichia coli	1000–2000
Aerobacter aerogenes	1000
Bacillus subtilis	500
Bacillus megatherium	500
Bacillus cereus var. *mycoides*	500

Table 17. Inhibitory action of sulfur dioxide against yeasts (Rehm et al. 1961, Rehm and Wittmann 1962)

Name of the test organism	pH value	Minimum inhibitory concentration tration of sodium sulfite in ppm
Saccharomyces cerevisiae	4.0	800–1600
Saccharomyces ellipsoideus	2.5–3.5	200– 800
Zygosaccharomyces nussbaumii	4.0	2000
Hansenula anomala	5.0	2400

Table 18. Inhibitory action of sulfur dioxide against molds (Rehm et al. 1961, Rehm and Wittmann 1962)

Name of the test organism	pH value	Minimum inhibitory concentration of sodium sulfite in ppm
Mucor species	2.5–3.5	300– 600
Penicillium glaucum	4.5	2800
Penicillium species	5.0	1600–4000
Penicillium species	2.5–3.5	200– 600
Aspergillus niger	4.5	2200

Owing to its relatively good antibacterial effect, sulfur dioxide is frequently combined with the preservatives sorbic acid and benzoic acid, which have more of a fungistatic action. This improves the antimicrobial spectrum of both groups of preservatives.

No notable instances of acquired resistance to sulfur dioxide by microorganisms other than in exceptional cases by *Zygosaccharomyces bailli* are known, although experience has shown that yeast strains cultivated in the absence of SO_2 are far more sensitive to this preservative than those constantly cultivated in media containing sulfur dioxide.

12.10 Fields of Use

There are many sectors of food technology where sulfur dioxide is employed partly or exclusively for reasons other than its antimicrobial effects. The following particulars are intended, however, to cover only the actions of sulfur dioxide on microorganisms.

12.10.1 Meat Products

Sulfites inhibit the development of bacteria on fresh meat and meat products (Kidney 1974). At the same time, sulfur dioxide stabilizes the color of meat to a certain extent and may thereby give the consumer a false impression as to its freshness. Consequently, the use of sulfur dioxide for meat is now classed in most countries as falsification and deceptive practice.

12.10.2 Fruit Products

Sulfur dioxide is used in fruit products as a temporary preservative. It is added primarily to raw or semi-finished products and removed again in the course of processing by the action of heat or vacuum so that the residual quantities in the end product are, on the whole, small.

Typical fields of use in which the antimicrobial action of the sulfur dioxide is an important factor are whole fruit or fruit portions for further processing, dried fruit, fruit juice, fruit juice concentrates, fruit pulps and fruit purees.

The applied concentrations of sulfur dioxide in fruit products are governed not only by the microbiological requirements. In nearly all cases sulfur dioxide has other technological functions to perform, such as protection against oxidative, enzymatic and non-enzymatic browning reactions and inhibition of chemically induced color and vitamin losses. The applied concentrations of sulfur dioxide necessary for this purpose are often higher than the concentrations needed to confer microbiological keeping properties. Depending on the product, between 0.01 and 0.2% SO_2 is employed in practice, or even higher concentrations may be used in special cases. The residual quantities of SO_2 remaining in the end-products are rarely higher than 0.01% and usually far lower. Such concentrations have only a slight antimicrobial action, if any at all, especially since some of the sulfur dioxide is still linked to ingredients of the fruit products specific to food, e.g. sugar.

12.10.3 Drinks

Among the drinks for which sulfur dioxide may be used, the most important is wine, including its primary products (Rose 1993).

The first use of sulfur dioxide is in the grape must. Additions of sulfur dioxide to the fresh must will prevent the growth of acetic acid bacteria, wild yeasts and molds. Sulfiting carried out properly will not damage the selected yeasts. Consequently, the addition of sulfur dioxide to grape must makes for steady and clean fermentation, besides inhibiting the development of acid-degrading bacteria and thus preventing premature acid degradation. Musts obtained at normal temperatures with a low acid content require some 40 to 50 mg SO_2/liter. In the case of musts rich in acids, 30 to 40 mg are sufficient, although those obtained at higher temperatures, e.g. in southern European countries, need up to 200 mg SO_2 per liter must.

Higher additions of 1,500 to 2,000 mg SO_2/liter completely prevent any further fermentation. Must given this radical treatment can be freed from the sulfur dioxide down to a residual quantity of about 25 to 150 mg/liter by heating to 90 to 110 °C in apparatus specially designed for the purpose, while inert gases are passed through at the same time (Wucherpfennig et al. 1973). The musts from which the sulfur is removed in this manner can be used for producing wines with "residual sweetness". The addition of sulfur dioxide or sulfites during fermentation as a means of halting fermentation is today regarded as undesired, since it involves a number of disadvantages, one being that the addition of SO_2 at this stage leads to a very high content of sulfur dioxide in the end-product.

During and after finishing of the wine, the added sulfur dioxide serves to remove acetaldehyde by chemical combination (which will not be discussed here in more detail), to stabilize the color, to obtain the correct redox potential and to achieve microbiological stability. Some of the sulfur dioxide becomes linked to various constituents of the wine and fermentation by-products, especially the acetaldehyde. The antimicrobial action of the sulfur dioxide is determined primarily by the unlinked portion, the so-called free sulfur dioxide, although the combined sulfur dioxide does retain some effect, at least against certain bacteria (Fornachon 1963, Lafon-Lafourcade and Peynaud 1974).

In accordance with its spectrum of action, the main effect of sulfur dioxide is to prevent bacterially induced changes in the wine, namely so-called wine diseases such as acescence, sweet wine disease, mannitic fermentation, mousiness and graisse. However, the SO_2 concentrations usual in wine making fail to prevent undesired growth of yeasts, also known as secondary fermentation. Some strains of yeast will continue to grow even in the presence of 1,000 mg SO_2 per liter wine (Schanderl 1959); so in order to stabilize wine with residual sweetness, use is nowadays made of sorbic acid, whose spectrum of action very favorably complements that of sulfur dioxide.

12.10.4 Other Points

Sulfur dioxide has been used since olden times in the form of 1–2% aqueous solutions as a disinfectant for equipment, barrels, bottles, corks and other ancillary items for the wine, drinks and food industries. By subsequently rinsing these in microbiologically untainted water and allowing them to drip dry, the amount of undesired SO_2 that finds its way into the final product is reduced to a minimum. Corks are adversely affected by long-term exposure to sulfur dioxide. Barrels are disinfected by fumigation with sulfur dioxide gas, which is produced by burning elemental sulfur inside the barrel. The sulfur dioxide formed has a disinfectant action.

12.11 Other Effects

Besides its preservative action, sulfur dioxide has a considerable number of other effects, some beneficial and others deleterious.

The most obvious drawback of sulfur dioxide is its intense, pungent odor, which may be apparent in foods it is used to treat. Consequently, sulfur dioxide is used preferably for preserving food intended for further processing.

Owing to its high chemical reactivity, sulfur dioxide may be involved in a variety of interactions with ingredients of food. Some of these interactions are undesired, whereas others are used technologically to good effect. The action of destroying thiamine is significant (Hölzel et al. 1966, Rehm and Wittmann 1962), although associated with high sulfur dioxide concentrations and low pH value. In addition, foods rich in vitamin B are rarely preserved with sulfur dioxide; so the practical repercussions of thiamine decomposition by SO_2 are kept within limits. Sulfur dioxide greatly reduces the decomposition of vitamin C in foods.

The chemical linkage of aldehydes to sulfur dioxide is utilized in wine making. This action is the main reason why sulfur dioxide is essential in the production of wine, because otherwise the fermentation by-product acetaldehyde would give the wine an undesired odor and flavor.

The reductive and antioxidative actions of sulfur dioxide are also very important for a number of branches of food technology. Enzymatic browning reactions are held in check by SO_2, namely by inhibiting the enzymes or by capturing the radicals that stimulate the process. Many non-enzymatic browning reactions, including the Maillard reaction, are also inhibited by sulfur dioxide (Sapers 1993).

Under certain conditions some sulfites may destroy aflatoxins formed in foods (Tabata et al. 1994).

Owing to misgivings concerning the toxicology of sulfur dioxide and sulfites a search for alternatives has been continuing for a long time. Possibilities discussed include 5,6-sulfinyl-L-ascorbic acid (Reutimann et al. 1991), ascorbic acid-2-phosphate (Sapers and Miller 1992), ascorbic acid itself, inhibitors for polyphenol oxidases and sulfur-containing amino acids (Sapers 1993). Like nitrite, sulfur dioxide is a multifunctional substance in which a large number of effects are combined. Some of the effects can be achieved with other preservatives, but so far no other substance has been found which also has the other effects of sulfur dioxide, such as its enzyme-inhibiting, reductive and antioxidative action.

12.12 General Literature

Green LF (1976) Sulphur dioxide and food preservation – a review. Food Chem 1, 103–124

Heydenreich GA (1967) Die schweflige Säure und ihre Salze in der Lebensmittelverarbeitung und -lagerung. Z Ernährungswiss 8, 44–65

Ough CS (1993) Sulfur dioxide and sulfites. In: Branen AL, Davidson PM: Antimicrobials in foods. Marcel Dekker, New York, p. 137–190

Roberts AC, McWeeny DJ (1972) The uses of sulphur dioxide in the food industry. A review. J Food Technol 7, 221–238

Wedzicha BL (1984) Chemistry of sulphur dioxide in foods. Elsevier Applied Science Publishers, London

12.13 Specialized Literature

Bassermann-Jordan F von (1923) Geschichte des Weinbaus. 2nd edition. Volume I. Frankfurter Verlags-Anstalt., Frankfurt p. 460–465

Belchi-Hernandez J, Florido-Lopez J, Estrada-Rodriguez J, Martinez-Alzamora F, Lopez-Serrano C, Ojeda-Casas J (1993) Sulfite-induced urticaria. Ann Allergy 71, 230–232

Bendtsen A, Jørgensen S (1994) Determination of total and free sulfite in unstabilized beer by flow injection analysis. J AOAC Intern 77, 948–952

Bhagat B, Lockett MF (1960) The absorption and elimination of metabisulphite and thiosulphate by rats. J Pharma Pharmacol 12, 690–694

Bhagat B, Lockett MF (1964) The effect of sulphite in solid diets on the growth of rats. Food Cosmet Toxicol 2, 1–13

Cabré F, Cascante M, Canela E (1990) A sensitive enzymatic method of sulfite determination. Anal Lett 23, 23–30

Cardwell T, Cattrall R, Chen G, Scollary G, Hamilton I (1993) Determination of sulfur dioxide in wines and beverages by flow injection analysis with reductive amperometric detection and electrolytic cleanup. J AOAC Intern 76, 1389–1393

Carr JG, Davies PA, Sparks AH (1976) The toxicity of sulphur dioxide towards certain lactic acid bacteria from fermented apple juice. J Appl Bacteriol 40, 201–212

Cluzan R, Causeret J, Hugot D (1965) Le métabisulfite de potassium. Étude de toxicité à long terme sur le rat. Ann Biol Anim Biochim Biophys 5, 267–281

Fazio T, Warner C (1990) A review of sulphites in foods: analytical methodology and reported findings. Food Add Contam 7, 433–454
Fitzhugh OG, Knudsen LF, Nelson AA (1946) The chronic toxicity of sulfites. J Pharmacol Exp Ther 86, 37–48
Fornachon JCM (1963) Inhibition of certain lactic acid bacteria by free and bound sulphur dioxide. J Sci Food Agric 14, 857–862
Gunnison AF (1981) Sulphite toxicity: A critical review of in vitro and in vivo data. Food Cosmet Toxicol 19, 667–682
Hayatsu H, Miura A (1970) The mutagenic action of sodium bisulfite. Biochim Biophys Acta 39, 156–160
Hötzel D, Muskat E, Cremer HD (1966) Toxizität von schwefliger Säure in Abhängigkeit von Bindungsform und Thiaminversorgung. Z Lebensm Unters Forsch 130, 25–31
Hoppe JO, Goble FC (1951) The intravenous toxicity of sodium bisulfite. J Pharmacol Exp Ther 101, 101–106
Hugot D, Causeret J, Leclerc J (1965) Effets de l'ingestion de sulfites sur l'excrétion du calcium chez le rat. Ann Biol Anim Biochim Biophys 5, 53–59
IARC (1992) IARC Monographs on the evaluation of carcinogenic risk to humans, Vol 54. IARC, Lyon, p. 131–186
Jaulmes P (1970) État actuel des techniques pour le remplacement de l'anhydride sulfureux. Bull OIV 43, 1320–1333
Kawamura Y, Kubo N, Arata H (1994) A microbial sensor for determination of sulfite in wines. J AOAC Intern 77, 1052–1058
Kidney AJ (1974) The use of sulphite in meat processing. Chem Ind (London) p. 717–718
King A, Ponting J, Sanschuck D, Jackson R, Mihara K (1981) Factors affecting death of yeast by sulphur dioxide. J Food Protect 44, 92–97
Lafon-Lafourcade S, Peynaud E (1974) Sur l'action antibacterienne de l'anhydride sulfureux sous forme libre et sous forme combinée. Connaiss Vigne Vin 8, 187–203
Lang K (1960) Die physiologischen Wirkungen von schwefliger Säure. B. Behr's Verlag, Hamburg
Lauteaume M-T, Ramel P, Girard P, Jaulmes P, Gasq M, Ranau J (1965) Effets physiologiques à long terme de l'anhydride sulfureux ou des sulfites utilisés pour le traitement des vins rouges. (Observations portant sur quatre générations de rats). Ann Falsif Expert Chim 58, 16–31
Lauteaume M-T, Ramel P, Jaulmes P, Manin D (1969) Détermination et comparaison des DL 50 du métabisulfite de potassium, de l'éthanal et de leur combinaison (hydroxy-éthane-sulfonate de potassium) par voie orale sur le rat de souche Wistar. Ann Falsif Expert Chim 62, 231–241
Lockett MF, Natoff IL (1960) A study of the toxicity of sulphite. I. J Pharm Pharmacol 12, 488–496
Monier-Williams GW (1927) Determination of sulphur dioxide in foods. Analyst 52, 415–416
Mukai F, Hawryluk I, Shapiro R (1970) The mutagenic specifity of sodium bisulfite. Biochem Biophys Res Commun 39, 983–988
Nagy S, Teuber S, Loscutoff S, Murphy P (1995) Clustered outbreak of adverse reactions to a salsa containing high levels of sulfites. J Food Protect 58, 95–97
Paíno-Campa G, Peña-Egido J, García-Moreno C (1991) Liquid chromatographic determination of free and total sulphites in fresh sausages. J Sci Food Agric 56, 85–93
Pfleiderer G, Jeckel D, Wieland T (1956) Über die Einwirkung von Sulfit auf einige DPN hydrierende Enzyme. Biochem Z 328, 187–194
Rehm H-J, Wittmann H, Stahl U (1961) Untersuchungen zur Wirkung von Konservierungsmittelkombinationen. VI. Das antimikrobielle Spektrum bei Kombinationen von Konservierungsmitteln. Z Lebensm Unters Forsch 115, 244–262
Rehm H-J, Wittmann H (1962) Beitrag zur Kenntnis der antimikrobiellen Wirkung der schwefligen Säure. I. Mitteilung. Übersicht über einflußnehmende Faktoren auf die antimikrobielle Wirkung der schwefligen Säure. Z Lebensm Unters Forsch 118, 413–429
Rehm H-J, Wittmann H (1963) Beitrag zur Kenntnis der antimikrobiellen Wirkung der schwefligen Säure. II. Mitteilung. Die Wirkung der dissoziierten und undissoziierten Säure auf verschiedene Mikroorganismen. Z Lebensm Unters Forsch 118, 465–478

Rehm H-J, Sening E, Wittmann H, Wallnöfer P (1964) Beitrag zur Kenntnis der antimikrobiellen Wirkung der schwefligen Säure. III. Mitteilung. Aufhebung der antimikrobiellen Wirkung durch Bildung von Sulfonaten. Z Lebensm Unters Forsch 123, 425–432

Rehm H-J, Wallnöfer P, Keskin H (1965) Beitrag zur Kenntnis der antimikrobiellen Wirkung der schwefligen Säure. IV. Mitteilung. Dissoziation und antimikrobielle Wirkung einiger Sulfonate. Z Lebensm Unters Forsch 127, 72–85

Reith JF, Willems JJL (1958) Über die Bestimmung der Schwefligen Säure in Lebensmitteln. Z Lebensm Unters Forsch 108, 270–280

Reutimann U, Bill R, Dürr P, Schobinger U (1991) Ersatz von SO_2 durch 5,6-Sulfinyl-L-Ascorbinsäure bei der Weinbereitung. Mitt Klosterneuburg 41, 65–70

Rose A (1993) Sulphur dioxide and other preservatives. J Wine Res 4, 43–47

Sapers GM (1993) Scientific status summary: Browning of foods: Control by sulfites, antioxidants and other means. Food Technol 47:10, 75–84

Sapers GM, Miller R (1992) Enzymatic browning control in potato with ascorbic-2-phosphates. J Food Sci 57, 1132–1135

Schanderl H (1956) Die Schweflige Säure des Weines in hygienischer Sicht. Z Lebensm Unters Forsch 103, 379–386

Schanderl H (1959) Die Mikrobiologie des Mostes und Weines. Stuttgart: Eugen Ulmer p. 244

Shtenberg AJ, Ignat'ev AD (1970) Toxicological evaluation of some combinations of food preservatives. Food Cosmet Toxicol 8, 369–380

Simon R, Stevenson D (1988) Adverse reactions to sulfites. In: Middleton et al. (eds) Allergy, Principles and Practice. Chapter 67, 3rd edition (vol. II). Mosby, St. Louis

Strahlmann B (1974) Entdeckungsgeschichte antimikrobieller Konservierungsstoffe für Lebensmittel. Mitt Geb Lebensmittelunters Hyg 65, 96–130

Sullivan J, Hollingworth T, Wekell M, Meo V, Saba H, Etemad-Moghadam A, Eklund C (1990) Determination of total sulfite in shrimp, potatoes, dried pineapple and white wine by flow injection analysis: a collaborative study. J Assoc Off Anal Chem 73, 35–41

Tabata S, Kamimura H, Ibe A, Hashimoto H, Tamura Y (1994) Degradation of aflatoxins by food additives. J Food Protect 57, 42–47

Til HP, Feron VJ, Groot AP de (1972 a) The toxicity of sulphite. II. Short- and long-term feeding studies in pigs. Food Cosmet Toxicol 10, 291–310

Til HP, Feron VJ, Groot AP de (1972 b) The toxicity of sulphite. I. Long-term feeding and multigeneration studies in rats. Food Cosmet Toxicol 10, 463–473

Vena G, Foti C, Angelini G (1994) Sulfite contact allergy. Contact Derm 31, 172–175

Wallnöfer P, Rehm H-J (1965) Beitrag zur Kenntnis der antimikrobiellen Wirkung der schwefligen Säure. V. Mitteilung. Die Wirkung der schwefligen Säure auf den Stoffwechsel atmender und gärender Hefe- und Colizellen. Z Lebensm Unters Forsch 127, 195–206

Warner C, Daniels D, Fitzgerald M, Joe F, Diachenko G (1990) Determination of free and reversibly bound sulphite in foods by reversed-phase, ion-pairing high-performance liquid chromatography. Food Add Contam 7, 575–581K (1978) Zur Toxizität der schwefligen Säure. Weinberg Keller 25, 171–184

Wucherpfennig K, Frank I, Bretthauer G (1973) Über die Herstellung von Süßreserve durch Stumm- und Entschwefelung. Allg Dtsch Weinztg 109, 850–858

CHAPTER 13

Chlorine

13.1 Synonyms

German: Chlor. *French:* Chlore. *Italian:* Cloro. *Spanish:* Cloro. *Russian:* Хлор.

13.2 History

Chlorine and compounds that release chlorine, namely bleaching powder (chloride of lime), were employed as long ago as the beginning of the nineteenth century as disinfectants in medicine. The successes of Semmelweis around the year 1847 in combating childbed fever by disinfecting the hands and equipment with chlorine were particularly spectacular.

This knowledge led to the use of chlorine for disinfecting equipment used in the production and storage of food, as well as drinking water and process water. In the disinfection of drinking water, chlorine has increasingly replaced ozone, the latter having been in use longer but being more expensive.

13.3 Commercially Available Forms, Derivatives

Chlorine is used as such or in the form of derivatives that release chlorine, chiefly hypochlorites, and in special cases as sodium and potassium dichloroisocyanurate. Chlorine is available as a liquefied gas in pressure vessels. Chlorine dioxide has an antimicrobial action of its own.

13.4 Properties

Cl_2, molar mass 70.91, at room temperature and normal pressure a green gas whose odor is perceptible even in concentrations as low as 0.001 mg/l air. The density of chlorine is about 2.5 times that of air. At 0 °C chlorine can be liquefied under a pressure of 3.76 bar. The quantity that dissolves in one liter of water is 4.6 g at 0 °C and 2.3 g at 20 °C.

NaClO, sodium hypochlorite, molar mass 74.44, deliquescent, readily water-soluble crystals, also marketed as an aqueous solution (soda bleaching lye, Eau de Labarraque).

$Ca(ClO)_2$, calcium hypochlorite, molar mass 142.99, $CaOCl_2$, molar mass 126.99, $Mg(ClO)_2$, molar mass 127.22, colorless powders with an odor of chlorine.

ClO_2, chlorine dioxide, molar mass 67.45, yellow, readily water-soluble gas.

13.5 Analysis

Chlorine releases iodine from iodides. When liquids containing chlorine are added to potassium iodide starch solution, a blue coloration results. Solutions of *o*-toluidine turn yellow to orange after the addition of chlorine. This yellow coloration can be used for colorimetric evaluation. Rapid and semi-quantitative determination of chlorine is possible with special test tubes (Drägerwerk 1991).

13.6 Production

Chlorine is manufactured by the electrolysis of hydrochloric acid or alkali chloride solutions. Owing to the difficulty of handling chlorine dioxide, it must be manufactured from sodium chlorite solution and hydrochloric acid at its place of use.

13.7 Health Aspects

Elemental chlorine has a severe inflammatory action on the skin and mucous membranes. A chlorine content of 700 ppm in the air (i. e. 2 g/m^3) is rapidly lethal since the alveoli of the lungs are irreversibly damaged by the hydrochloric acid produced. Inhalation of air with a content of 20 ppm (i.e. 60 mg/m^3) for 15 minutes may have fatal consequences. The maximum allowable workplace concentration for elemental chlorine is 1.5 mg/m^3.

13.8 Regulatory Status

Chlorine and both inorganic and organic products that release chlorine such as sodium, calcium and magnesium hypochlorite, chloride of lime, chlorine dioxide and sodium and potassium dichloroisocyanurate are permitted in many countries for disinfecting drinking water. The maximum permissible quantities are 1.2 mg chlorine and hypochlorites, and 0.4 mg chlorine dioxide, per liter water.

13.9 Antimicrobial Action

13.9.1 General Criteria of Action

In the correct concentration chlorine kills microorganisms rapidly. Consequently, chlorine is a disinfectant rather than a preservative.

The antimicrobial action of chlorine is based on its powerful oxidizing effect and rapid linkage to proteins. Chlorine also attaches very easily to double bonds of biomolecules. These reactions have a lethal effect on the metabolism of the microorganism (Trueman 1971). In addition, chlorine destroys the cell membrane and reacts with the DNA (Dychdala 1977).

The action of chlorine is severely impaired by the presence of organic matter because this reacts with some of the active substance. The reaction of chlorine with ammonia and amino compounds also reduces the power of its effect. This chlorine loss has to be taken into account in practical conditions. Consequently in heavily polluted media such as sewage, the chlorine factor used in calculations is that for the active constituent, the "available free chlorine".

Heat considerably intensifies the antimicrobial action of chlorine. Bacillus spores at 50 °C, for instance, are killed ten times as rapidly as at 20 °C with the same chlorine concentration (Trueman 1971).

Chlorine acts best in the neutral to weakly acid pH range. At pH 6, for example, its action is between 2 and 60 times as rapid as at pH 10, depending on the type of microorganism (Trueman 1971).

13.9.2 Spectrum of Action

Chlorine has a very broad antimicrobial spectrum of action. It acts both against bacteria, including spores, yeasts and molds and against algae, protozoa and many viruses. A considerable number of bacteria can be killed under optimum conditions (low microbe count, pH 7, temperature of 20 to 25 °C) with as little as 0.05 to 0.1 ppm chlorine. Tubercle bacilli, salmonella, proteus and pseudomonas strains are somewhat more resistant. Molds are about ten times more resistant than bacteria (Trueman 1971, Dychdala 1977).

13.10 Fields of Use

13.10.1 Drinks

The most important field of use for chlorine is the disinfection of drinking water. Usually the "indirect chlorination process" is employed. This involves transferring the chlorine obtained from pressure vessels into a solution containing about 1 to 5 g chlorine per liter in suitable mixing devices and adding this solution to the drinking water to be treated. The use of hypochlorite solutions and other compounds that release chlorine is only of relatively minor importance.

A problem that arises in drinking water disinfection by chlorination is the haloform reaction. Elemental chlorine reacts with organic impurities in the drinking water, such as amino acids and humic substances, to form trihalomethanes, e.g. the carcinogen chloroform (IARC 1991). If chlorine dioxide is used instead of chlorine the scale of the haloform reaction can be reduced.

13.10.2
Other Points

Chlorine is also used for disinfecting containers and equipment for foods. Recently in the USA, chlorine and chlorine dioxide have been employed in the form of aqueous solutions for salmonella decontamination of poultry (Anon 1995). Their use is also documented in fruit-washing water (Roberts and Reymond 1994) and, in slow-release form, in polymer food packaging (Wellinghoff 1995).

13.11
Other Effects

Chlorine and compounds that release chlorine are powerful oxidizing agents and thus highly corrosive.

The high toxicity of the free chlorine should be borne in mind. Although the dosages employed in drinking water are considered harmless, their odor and flavor may be disturbing.

13.12
Literature

Anon (1995) Chlorine dioxide use in poultry preserving cleared by FDA. Food Chem News dated March 6, 1995, p. 53

Drägerwerk (1991) Dräger-Röhrchen Handbuch. Drägerwerk AG, Lübeck

Dychdala GR (1977) Chlorine and chlorine compounds. In: Block SS: Disinfection, Sterilization and Preservation. Lea and Febiger, Philadelphia, p. 319–324

IARC (1991): International Agency for Research and Treatment on Cancer, Monograph 52, Chlorinated drinking water. IARC, Lyon, p. 15–359

Roberts R, Reymond S (1994) Chlorine dioxide for reducing of postharvest pathogen inoculum during handling of tree fruits. Appl Environm Microbiol 60, 2864–2868

Trueman JR (1971) The halogens. In Hugo, WB: Inhibition and Destruction of the Microbial Cell. Academic Press, London, p. 137–183

Wellinghoff S (1995) Keeping food fresh longer. Technol Today June 1995, 20–22

Chapter 14

Ethanol

14.1 Synonyms

IUPAC: Ethanol
English: Ethyl alcohol, "alcohol". *German:* Ethylalkohol, Weingeist, "Alkohol". *French:* Alcool éthylique, ethanol, "alcool". *Italian:* Alcool etilico, etanolo, "alcool". *Spanish:* Alcohol etílico, ethanol, "alcohol". *Russian:* Этанол, этиловый спирт, "спирт".

14.2 History

Alcohol was already being used by the Arabs to preserve fruit a thousand years ago. An even older practice, which could likewise be termed a preserving process in a certain sense, is the fermentation of sugar-containing fruit into wine. In the industrial sector the preservation of food, and particularly fruit, in alcohol has never attained any major importance. This contrasts with the private household in some European countries, where the practice has retained its importance to the present day.

14.3 Properties

C_2H_5OH, molar mass 46.07, colorless liquid which boils at 78 °C and is miscible with water in any proportions. Together with water, ethanol forms an azeotropic mixture of 95.6% alcohol and 4.4% water.

14.4 Analysis

For the quantitative determination of alcohol, the food to be investigated is distilled and the alcohol content of the distillate is determined pycnometrically. Enzymatic determination with alcohol dehydrogenase (ADH) (Beutler and Michal 1977) and chemical determination by Widmark's method using oxidation with potassium dichromate solution are also suitable for determination of very low alcohol contents. Other quantitative determination methods for ethanol are gas chromatography and ion chromatography (Barka and Heidger 1989).

14.5 Production

The ethanol employed in the food sector is obtained exclusively by the fermentation of sugar-containing liquids.

14.6 Health Aspects

14.6.1 Acute Toxicity

The LD_{50} of ethanol after oral administration is 9.5 g for the mouse, 13.7 g for the rat, 6.3 to 9.5 g for the rabbit, and 5.5 to 6.5 g for the dog per kg body weight (Spector 1956). For the human adult, the quantity of alcohol hazardous to life when ingested within a brief period is about 200 to 400 ml, calculated as pure alcohol and representing a blood concentration of 4 to 6‰ (Widmark 1933). Persons habituated to alcohol have been known to survive blood alcohol levels as high as 12‰ in isolated cases.

14.6.2 Chronic Toxicity

Frequent and regular administration of alcohol leads to habituation. As a result, doses that would originally have had an intoxicating and narcotic effect cease to have a paralytic action on the central nervous system. Instead, symptoms of irritation occur in the gastro-intestinal tract and other digestive disorders arise, while in an advanced stage the condition will give rise to fatty liver and cirrhosis. The quantity of alcohol considered tolerable over the long term is 30 g for men and 15 g for women daily. Chronic ingestion of larger quantities may cause fatty liver and hence chronic hepatitis. Ethanol is not in itself a carcinogen, but under certain conditions it stimulates chemically induced carcinogenesis. Excessive ethanol consumption is a risk factor for the genesis of liver tumors as a result of Hepatitis B infection (Seitz et al. 1991, Seitz et al. 1993), the mamma (Toniolo et al. 1989) and the rectum (Wu et al. 1987). The action of ethanol is mutagenic (Shapero et al. 1978, Hayes 1985) and teratogenic (Obe and Ristow 1979).

The maximum allowable workplace concentration for ethanol is 1900 mg/m^3.

14.6.3 Biochemical Behavior

Alcohol can be absorbed not only by the gastric mucous membrane but also by the skin and lungs. It is excreted in small quantities (about 5%) by the lungs and in the urine, while most is oxidized through the intermediate stage of acetaldehyde to yield carbon dioxide and water. The rate at which alcohol is eliminated from the blood is about 15 mg per 100 ml per hour. The body is capable of metabolizing about 100 mg alcohol per hour per kg body weight.

14.7 Regulatory Status

As alcohol is an essential constituent of many foods, its addition to food for preservation purposes is not subject to any food law control. Exceptions to this are alcoholic drinks, e.g. wine, whose alcohol content in a number of countries must not be increased by adding extraneous alcohol, or else may be increased only in exceptional cases. In most countries ethanol is not considered as a food additive.

14.8 Antimicrobial Action

In high concentrations, ethanol acts against microorganisms, by denaturing proteins of the protoplasm. The most effective alcohol concentration is one of 60 to 70 %, this being the reason why such a concentration is usual when alcohol is employed as a disinfectant. Owing to its unspecific denaturing action on protein, alcohol is effective against microorganisms of all kinds. Bacteria are basically more sensitive than yeasts. In many cases they are inhibited by alcohol concentrations of less than 10 %. Only *Streptococcus faecalis* and some lactic acid bacteria tolerate higher alcohol concentrations (Yamamoto et al. 1984). Certain yeasts tolerate up to 20 % alcohol. In concentrations of around 50 %, ethanol inactivates all vegetative microorganisms within a short time, including fungal conidia, though bacterial spores are not damaged (Schmidt-Lorenz and Hotz 1985).

Lower applied concentrations of 5 to 20 % have a preserving action by lowering the water activity of food. If the water activity of an intermediate moisture food system is reduced by means of suitable humectants, preservative effects can be attained with as little as 2 to 4% added alcohol (Shapero et al. 1978). Ethanol enhances the antimicrobial action of sorbic acid (McCarthy et al. 1988, Parish and Carroll 1988).

14.9 Fields of Use

14.9.1 Fruit Products

One method of preservation employed in the home is the “Rumtopf”, a German specialty. This involves the sugaring of fruit, especially berries and drupes, which are then immersed in rum of high alcohol content. Brandy and other highly alcoholic spirits may be used as alternatives to rum. Since the fruit contain a great deal of water in their fresh state, and thus dilute the alcohol, the total alcohol content of the “Rumtopf” must not fall appreciably below 20% as otherwise the fruit might start to ferment. In the "Rumtopf" not only the alcohol but also the added sugar is important as a preservative.

When jams are prepared in the home, greaseproof paper saturated with alcohol is sometimes still placed on the jam, once transferred to the jars, to prevent mold growth.

14.9.2 Drinks

Fruit juices for the production of liqueurs are frequently fortified with alcohol. The juices used may be unfermented or fermented with alcohol contents of 4 to 5%.

In wine the danger of microbial spoilage declines with an increase in alcohol content. This applies to bacterially induced wine diseases as well as to secondary fermentation caused by yeasts. It is reasonable to assume that the alcohol occurring naturally in wine is, by itself, no reliable protection against microbial spoilage. Only when the alcohol is increased to a total content of 19 to 20% by adding alcohol distillates do the products become completely stable microbiologically. In most countries the addition of alcohol to wine is prohibited as a general principle and allowed only for special products in which its use is customary. Examples of these products are fortified wines such as sherry and port.

A further important use of alcohol is its addition to grape juice in order to prevent fermentation by yeast. Grape juice that has been muted i.e. rendered "unfermentable" by this method is known as mistelle and used chiefly for sweetening of fortified wine.

14.9.3 Baked Goods

In a number of countries where sorbic acid is not permitted for the purpose, ethanol is sometimes used to preserve cut bread, small cakes or part-baked dough. This is done by spraying high-percentage alcohol into the package shortly before packing and sealing. As a result the baked goods have alcohol concentrations between 0.5 and 2%. At these applied concentrations, ethanol also keeps the goods fresh by delaying the staling process (Seiler 1979). Ethanol may also be used for this purpose in combination with protective gases (see Sect. 7.10.6) (Vora et al. 1987).

14.10 Other Effects

Ethanol is enjoyed in alcoholic drinks. Because of its good dissolving properties it is also important as a solvent in extraction processes and as a carrier.

14.11 Literature

Barka G, Heidger V (1989) Erfahrungen mit der neuen Bestimmungsmethode für Zucker und Alkohole in der Lebensmittelanalytik. GIT Suppl (2) 32–38

Beutler H-O, Michal G (1977) Neue Methode zur enzymatischen Bestimmung von Äthanol in Lebensmitteln. Z Anal Chem 284, 113–117
Hayes S (1985) Ethanol-induced genotoxicity. Mutat Res 143, 23–27
McCarthy T, van Eeden A, Stephenson N, Newman C (1988) Interaction between ethanol and selected antimicrobial preservatives. S Afric J Sci 84, 128–132
Obe G, Ristow H (1979) Mutagenic, cancerogenic and teratogenic effects of alcohol. Mutat Res 65, 229–259
Parish M, Carroll D (1988) Minimum inhibitory concentration studies of antimicrobic combinations against *Saccharomyces cerevisiae* in a model broth system. J Food Sci 53, 237–242
Sandler RS (1983) Diet and cancer: Food additives, coffee and alcohol. Nutrition Cancer 4, 273–279
Schmidt-Lorenz W, Hotz F (1985) Ethanolvorbehandlung zur selektiven Koloniezählung von Bacillussporen bei Gewürzen. Mitt Gebiete Lebensm Hyg 76, 127–155
Seiler DA (1979) Flour Milling and Baking. Res Assoc Bull 2, 64
Seitz M, Heipertz W, Osswald B, Hörner M, Simanowski U, Egerer G, Kommerell B (1991) Leberschäden durch Alkohol. Bundesgesundheitsbl 39, 101–104
Seitz M, Simanowski U, Lieber C (1993) Alkoholismus und alkoholische Organschäden. Heidelberg: Hüthig M, Nelson DA, Labuza TP (1978) Ethanol inhibition of *Staphylococcus aureus* at limited water activity. J Food Sci 43, 1467–1469
Spector WS (1956) Handbook of Toxicology, Volume 1. WB Saunders, Philadelphia, p. 128–131
Toniolo P, Riboli E, Protta F, Charrel M, Coppa A (1989) Breast cancer and alcoholic consumption: A case-control study in northern Italy. Cancer Res 49, 5203–5206
Vora M, Sidhu J (1987) Effects of varying concentrations of ethyl alcohol and carbon dioxide on the shelf life of bread. Chem Mikrobiol Technol Lebensm 11, 56–59
Widmark EMP (1933) Die Maximalgrenzen der Alkoholkonsumption. Biochem Z 259, 285–293
Wu A, Paganini-Hill A, Ross B, Henderson B (1987) Alcohol, physical activity and other risk factors for colorectal cancer: A prospective study. Brit J Cancer 55, 687–694
Yamamoto Y, Higashi K, Yoshi H (1984) Inhibitory activity of ethanol on food spoilage bacteria. Part II. Studies on growth inhibition of food spoilage microorganisms for low salt foods. Nippon Shokuhin Kogyo Gakkaishi 31, 531–535

Chapter 15

Sucrose

15.1 Synonyms

IUPAC: Sucrose
English: Cane sugar, beet sugar, "sugar". *German:* Saccharose, Rohrzucker, Rübenzucker, "Zucker". *French:* Saccharose, sucre de canne, sucre colonial, sucre de betteraves, "sucre". *Italian:* Saccarosio, zucchero di canna, zucchero di barbabietola, "zucchero". *Spanish:* Sacarosa, azúcar de cana, azúcar de remolacha, "azúcar". Russian: Сахароза, тростниковый сахар, свекловичный сахар, "сахар".

15.2 History

It was from South-East Asia and in about A.D. 700 that sugar – in the form of cane sugar – first came to Europe, from where its cultivation later spread to America. The use of sugar beet as an industrial source of sugar is a German development dating from the eighteenth century. At first, sugar was purely a luxury commodity and medicine on account of its scarcity and cost, but in the course of time it was used increasingly for the sweetening of foods and later still for food preservation.

15.3 Commercially Available Forms

Sucrose is employed almost exclusively in the form of cane sugar, obtained from sugar cane, and beet sugar, obtained from sugar beet. It is used in solid form or as solutions (syrups). Compared with sucrose, the other sugar varieties are of only minor importance for preservation purposes.

15.4 Properties

$C_{12}H_{22}O_{11}$, molar mass 342.30, white, sweet-tasting crystals which melt at 185 °C. Sucrose is only slightly soluble in alcohol. Some 204 g sucrose dissolve in 100 g water at room temperature. A saturated sucrose solution contains 67.1 g per 100 g solution at room temperature. At 100 °C, 487 g sucrose will dissolve in 100 g water.

15.5 Analysis

Detection and quantitative determination of sucrose in foods can be carried out with enzymatic techniques (Gromes and Mörsberger 1992), polarimetry, thin-layer chromatography, gas chromatography, ion chromatography (Barka and Heidger 1989) and wet assay methods, e.g. Luff-Schoorl, as well as by HPLC (Schleich and Engelhardt 1989, Solbrig-Lebuhn 1992, Kaufmann 1993).

15.6 Production

Sucrose is obtained chiefly from sugar cane (*Saccharum officinarum*) or sugar beet (*Beta vulgaris saccharifera*). In both plants the sucrose is available as such. This is extracted from the corresponding parts of the plants with water and separated out by crystallization after the solutions have been purified.

15.7 Health Aspects

The LD_{50} of sucrose determined from feeding to rats is in the region of 30–35 g/kg body weight, the male rats displaying a slightly more sensitive reaction than the females (Boyd et al. 1963).

The action of sugar in causing caries is undisputed (Schraitle and Siebert 1987). The fermenting properties are crucial in determining the cariogenic potential of the different sugar varieties (DGE 1988). The development of caries is determined not so much by the amount of sugar consumed per day, but the length of time for which it remains in the oral cavity and the frequency of consumption. Sugar clearance is much faster for drinks than for solid foods, e.g. confectionery.

Like any other high-calorie food, sugar may help to cause obesity. This may promote the onset of certain forms of diabetes, even though sugar may not be the actual cause of the disease (WHO 1990). The WHO (WHO 1990) and the DGE (DGE 1988) regard the upper sucrose consumption limit as being the quantity which represents 10 % of an individual's daily energy consumption.

Claims that sugar encourages disease, for instance by reducing vitamin B_1 levels, are not supported by scientific evidence (Vettorazzi and MacDonald 1988, Bässler 1991, Black 1993).

15.8 Regulatory Status

As sugar is a major ingredient in many foods, its addition for preservation purposes is not subject to any food law restrictions. Quantitative restrictions exist only for its use in dietary products for consumption by diabetics or the obese, as well as for fruit products for reasons of quality. In most countries, sugar is not defined as an additive.

15.9 Antimicrobial Action

15.9.1 General Criteria of Action

Sucrose lowers the water activity (a_w value) of a system and thus reduces the opportunities for microorganisms to live. Its effect is therefore very similar to that of common salt (see Sect. 6.9.1). Since the a_w value of saturated sugar solutions is only around 0.85 and some microorganisms can grow even below this limit, foods cannot be protected against microbial contamination with sugar alone.

The foods to be preserved can be immersed in solutions containing greater or lesser quantities of sugar; alternatively, sugar can be added in dry form. Osmotic action then causes water to be extracted from the food, which adjusts to a certain water activity, its level depending on the sugar content. The corresponding relationships are given in Table 19. The combination of the sugar with physical preservation methods, especially concentration, drying and heating, is of great practical importance, as also is the combination of sucrose with other preservatives (Romming and Frank 1988). Sucrose has no direct antimicrobial action; indeed, in small concentrations it is actually a nutrient for many microorganisms. As a result of osmotic effects, sucrose increases the heat resistance of molds (Doyle and Marth 1975).

15.9.2 Spectrum of action

As sucrose acts by reducing the water activity, the demands imposed by the various microorganisms on the water activity determine the spectrum of action of the sucrose. Table 10 shows the limiting values for some important microorganisms. The microbes most easily inhibited are the bacteria.

Among the microorganisms that tolerate especially high sugar concentrations, mention should be made of *Aspergillus glaucus* and the so-called osmotolerant

Table 19. Water activity of sucrose solutions (Seiler 1969)

a_w value	Content of the solution in g sucrose/100 g H_2O
0.99	11
0.96	25
0.95	78
0.94	93
0.93	107
0.92	120
0.90	144
0.88	169
0.86	194
0.85	208

yeasts, e.g. *Zygosaccharomyces rouxii, bailii* and yeasts of the genus *Torulopsis*. Some of these are not only osmotolerant but also saccharophilic, i.e. they not only tolerate high sugar concentrations but also grow better in the presence of large sugar quantities. Thus *Zygosaccharomyces rouxii*, for example, grows better in 30% sugar solutions than in sugar-free or 60% sugar solutions (Bills et al. 1982).

Like common salt, sucrose lowers the solubility of oxygen in water. Hence, in products containing a high proportion of sugar the oxygen available to aerobic microorganisms is only a fraction of that found in products with a low sugar content (Lubieniecki-von Schelhorn 1975).

15.10 Fields of Use

15.10.1 Fruit Products

When preserving fruit products with sugar, a distinction is drawn between two types of process; one involves simultaneous concentration and the other merely comprises an addition of sugar. Depending on the product to be preserved, the sugar is used either in solid form or as a solution (syrup).

Among the products preserved by sugar and involving simultaneous concentration, the most important are jams and marmalades. Depending on the nature of the product, the sucrose content is between 55 and 65%, and in the case of marmalades even higher.

Products preserved by the addition of sugar, such as jams and marmalades, have water activities of 0.75 to 0.82. Such water activity levels are not always low enough to suppress completely any mold growth and fermentation by osmotolerant yeasts. For this reason it is necessary in many instances to include a pasteurization process or to use other preservatives as well, e.g. sorbic acid, in the form of preserving sugar. This is a mixture of sucrose, pectins and sorbic acid.

The addition of sugar without concentration is used for preserving candied fruit, lemon peel and orange peel. These products are obtained from whole fruit or fruit portions simply by treatment with concentrated sugar solutions. In this process, much as with salting and pickling, osmosis brings about an exchange between the concentrated sugar solution and the juice in the cells of the fruit. In general, the fruit to be preserved are immersed in sugar solutions of ever-increasing concentration. The final products have a higher sugar content than jams and marmalades and consequently can usually be preserved without additional preservation processes or preservatives. The same applies in the case of fruit syrups; these are prepared by adding sucrose to fruit juices and contain up to 68% sugar.

15.10.2 Baked Goods

Even though sucrose is added to many baked goods primarily for reasons of taste, it is frequently also the crucial factor in their preservation. Typical examples of the use of sugar as a preservative are cakes and other fine baked goods. Depending on their type, their sugar content is in the region of 20 to 50%. Such additions of sugar produce water activities of only 0.83 to 0.90, which are frequently inadequate for reliable preservation over a lengthy period; so occasionally other preservatives such as sorbic acid are added as well.

15.10.3 Confectionery

Some varieties of confectionery use sucrose almost exclusively to provide body and flavor, whereas in others sucrose also assumes the function of a preservative for other ingredients. The second group includes such products as marzipan, nougat, and fillings for chocolates and pralines. In these, the additions of sucrose often exceed 60%. Together with the dry matter content of the other ingredients it is generally sufficient to eliminate microbial spoilage. Only in isolated cases is it necessary to combine the preservative, sucrose, with other more powerful preservatives.

15.11 Other Effects

Besides its preservative action, sucrose can provide a markedly sweet taste, which is actually the main reason in many cases for using sugar as a food ingredient. The sugar concentrations producing the best taste are much lower than those necessary for preserving foods. In the lower concentration range, mainly in additions below 10%, sucrose has no preserving properties and in fact serves as a nutrient for many microorganisms, either directly or after conversion into glucose and fructose.

15.12 Literature

Bässler K (1990) Zucker: Ernährungsmedizinische Bedeutung von Zucker – Eine Bestandsaufnahme. Z Ernährungswiss 29, Suppl 1, 1–69

Barka G, Heidger V (1989) Neue Bestimmungsmethode für Zucker and Alkohol in Getränken. GIT Suppl (2), 32–38

Bills S, Restaino L, Lenovich LM (1982) Growth response of an osmotolerant sorbate-resistant yeast, *Saccharomyces rouxii*, at different sucrose and sorbate levels. J Food Protect 45, 1120–1124

Black P (1993) Sucrose in health and nutrition – Facts and myths. Food Technol 47 (1), 130–133

Boyd EM, Godi I, Abel M (1963) Acute oral toxicity of sucrose. Toxicol Appl Pharmacol 7, 609–618

DGE (1988) Ernährungsbericht. Henrich, Frankfurt, p. 34

Doyle MP, Marth EH (1975) Thermal inactivation of conidia from *Aspergillus flavus* and *Aspergillus parasiticus*. II. Effects of pH and buffers, glucose, sucrose and sodium chloride. J Milk Food Technol 38, 750–758

Gromes R, Mörsberger F (1992) Enzymatische Zuckerbestimmung. Bio Tec 2, 32

ICUMSA (1990) Internat Comm for Uniform Methods of Sugar Analysis. 20th Session. British Sugar Technical Centre, Norwich

Kaufmann A (1993) Übersicht und Anwendung der HPLC-Analytik von Säuren und Zucker im Wein. Mitt Geb Lebensmittelunters Hyg 84, 311–333

Lubieniecki-von Schelhorn M (1975) Die Sauerstoffkonzentration als bestimmender Faktor für mikrobielle Vorgänge in Lebensmitteln unter besonderer Berücksichtigung einer sauerstofffreien Verpackung. Verpack-Rundsch 26, Wiss Beilage zu Nr 1, p. 1–6

Romming I, Frank H (1988) Growth response of putrefactive/anaerobe 3679 to combinations of potassium sorbate and some common curing ingredients (sucrose, salt and nitrite), and to noninhibitory levels of sorbic acid, J Food Protect 51, 651–654W, Engelhardt H (1989) Möglichkeiten der HPLC in der Zuckeranalytik. GIT Fachz Lab 33, 624, 630

Schraitle S, Siebert G (1987) Zahngesundheit und Ernährung. Hanser, München

Seiler DAL (1969) Equilibrium relative humidity of baked products with particular reference to the shelf life of cakes. In: BFMIRA: Relative humidity in the food industry. Proceedings of a symposium held in London. 16.9.1969. Symposium Proc No 4, p. 28–36

Solbrig-Lebuhn M (1992) Kohlenhydratanalytik. Laborpraxis 16, 786–789

Vettorazzi G, MacDonald I (1988) Sucrose. Nutritional and safety aspects. Springer, Berlin Heidelberg New York

WHO (1990) Diet, nutrition and the prevention of chronic diseases. Report of a WHO study group. Techn Rep Ser No 797. Geneva: WHO

Chapter 16

Formic Acid

16.1 Synonyms

IUPAC: Methanoic acid
German: Ameisensäure. *French:* Acide formique. *Italian:* Acido formico. *Spanish:* Acido fórmico. *Russian:* Муравьиная кислота.

16.2 History

The preservative action of formic acid has been known for over 100 years (Jodin 1865). Since then formic acid has been used in a number of countries, primarily in central and eastern Europe, as a food preservative, chiefly for fruit products. Recently, however, formic acid has been replaced increasingly by other preservatives or preservation processes and is consequently now of only minor importance.

16.3 Commercially Available Forms, Derivatives

For food preservation purposes, formic acid is employed in the form of aqueous solutions of the acid itself or in the form of sodium, potassium or calcium formate.

16.4 Properties

HCOOH, molar mass 46.03, colorless, transparent liquid with a pungent odor, unreservedly miscible with water, solidifies in pure form at 8 °C and boils at 101 °C.

Sodium formate, molar mass 68.01, potassium formate, molar mass 84.13, calcium formate, molar mass 130.11, white crystalline powders, more or less readily water-soluble.

16.5 Analysis

For qualitative detection and quantitative determination of formic acid in foods, the acid has to be isolated beforehand from the product to be investigated, e.g. by acid steam distillation. Formic acid can then be readily reduced to formaldehyde by heating in an acid solution with metallic magnesium. Together with chromotropic acid (1,8 dioxynaphthalene 3,5-disulfonic acid) this yields a violet-colored solution (Bremanis 1949). The reaction can also be employed for the quantitative determination of the formic acid. Ethyl formate, produced when formic acid is reacted with ethanol, is converted by hydroxylamine to formylhydroxamate, which reacts with iron (III) salts to give red iron hydroxamate, the color of which can be evaluated photometrically (Tanner 1976). Formic acid can also be detected by gas chromatography or enzymatic determination (Reinefeld and Bliesner 1977).

16.6 Production

Formic acid occurs as a by-product in a number of large-scale chemical processes, for example in the manufacture of acetic acid by the oxidation of hydrocarbons. Planned methods for producing formic acid include that of hydratizing CO or the saponification of formamide, which can be produced beforehand from CO, methanol and ammonia. Some of these processes yield formates, from which the formic acid can be released by mineral acids. Otherwise formates are obtained by reacting formic acid with the corresponding metal hydroxides.

16.7 Health Aspects

16.7.1 Acute Toxicity

According to available data, the acute toxicity of formic acid for rodents ranges from 1.1 to 1.8 g/kg body weight (Malorny 1969 a, Sporn et al. 1962). Potassium formate and sodium formate are considerably less toxic, with an LD_{50} of 3 and 7.4 g/kg body weight respectively (Malorny 1969 a). Formic acid is lethal to humans in a quantity of 50 to 60 g (Tracor-Jitco 1974).

In high concentrations, formic acid irritates the skin and mucous membranes and must therefore be handled with care.

16.7.2 Subchronic Toxicity

Additions of 0.5 to 1% formic acid to the drinking water of rats cause disturbances to the rats' growth and damage to organs (Sporn et al. 1962). In humans a daily intake of 0.5 g formic acid, i.e. 8 mg/kg body weight, over a period of 4 weeks had no noticeable effect (Life Sciences Research Office 1976).

16.7.3 Chronic Toxicity

When 1% formic acid is added to the drinking water of rats for seven months, changes occur in the rats' blood picture and the survival rate of the young is reduced (Sporn et al. 1962). In a two-year feeding test in which rats consumed drinking water containing 0.2% calcium formate ad libitum, representing 150 to 200 mg calcium formate per kg body weight, no damage specifically due to the calcium formate occurred to the function of the organs. Moreover, there was no untoward effect on the growth, fertility or development of the animals in the next generations (up to 5). In a further experiment, in which double the dosage was administered to two generations, the results were the same (Malorny 1969 a).

Formic acid has a mutagenic effect on insects, e.g. *Drosophila*, and some microorganisms, but not on mammal cells (Tracor-Jitco 1974), (Life Sciences Research Office 1976).

Formic acid and formates are not teratogenic (Malorny 1969), (Tracor-Jitco 1974), (Life Sciences Research Office 1976), nor are they carcinogenic (Tracor-Jitco 1974).

16.7.4
Biochemical Behavior

Formic acid is readily absorbed by the body on account of its good solubility. It can also be absorbed though the skin and mucous membranes (Malorny 1969 b). In addition it is a normal constituent of human blood and tissue and plays an important part in the intermediate metabolism in the transfer of C_1 substances (Malorny 1969 a, Palese and Tephly 1975). A small quantity of ingested formic acid is excreted in unchanged form in the urine, while the majority is metabolized. In humans to whom sodium formate is administered orally, the biological half-life for formic acid in the blood plasma is in the region of 45 minutes (Malory 1969 b).

16.8
Regulatory Status

Formic acid, sodium formate and calcium formate were permitted in some European countries for the preservation of foods such as fish products, some fruit products intended for further processing and pickled vegetables. Under future EU legislation their use will no longer be permitted.

16.9
Antimicrobial Action

16.9.1
General Criteria of Action

Part of the action of formic acid, especially that against bacteria, is based on its effect in lowering the pH. In common with acetic and lactic acid, free formic acid differs from higher-molecular organic acids in that it acts only in relatively high concentrations which appreciably reduce the pH of the medium to be preserved. In addition, formic acid has an antimicrobial effect independent of the action of the acid as such. Even at pH values where it exists in dissociated form in appreciable quantities it still inhibits decarboxylases (Scheer 1971 a) and hemin enzymes (Lück 1958), especially catalase, the latter to an unusual degree in comparison with other acid preservatives (Lück 1957).

Formic acid is the strongest of the aliphatic carboxylic acids and because of its high dissociation constant is suitable only for the preservation of highly acid products in the pH range below 3.5. Formates have no antimicrobial action in the weakly acid to neutral pH range (Scheer 1971 b); the action is acquired only when acid is added (Scheer 1971 a).

Formic acid increases resistance phenomena to a greater extent than other aliphatic carboxylic acids, e.g. in the case of *Escherichia coli* after sublethal exposure (Lück and Rickerl 1959), although the viability of the strains with increased resistance is only limited (Rehm 1961 a).

16.9.2 Spectrum of Action

The action of formic action is directed principally against yeasts and a number of bacteria (see Tables 20 to 22). Lactic acid bacteria and molds are fairly resistant (Rehm 1961 b, Balatsouras and Polymenacos 1963).

Table 20. Inhibitory action of formic acid against bacteria (Rehm 1961 b)

Name of test organism	pH value	minimum inhibitory concentration in ppm
Pseudomonas species	6.0	2500– 7500
Micrococcus pyogenes	6.0	14000
Streptococcus lactis	5.2	4000
Lactobacillus arabinosus	6.0	35000–50000
Escherichia coli	5.2	700– 1000
Bacillus species	6.0	500– 5000

Table 21. Inhibitory action of formic acid against yeasts (Rehm 1961 b, Balatsouras and Polymenacos 1963)

Name of test organism	pH value	minimum inhibitory concentration in ppm
Saccharomyces species	4.0–4.5	1200–10000
Saccharomyces species	5.0	800–36000
Asporogenic yeasts	5.0	1000–36000
Trichosporon variabile	3.9	1000
Hansenula subpelliculosa	3.0	10000–12500
Hansenula anomala	3.0	8000
Candida krusei	3.0	16000
Torulopsis stellata	3.0	6000–16000
Oospora lactis	3.0	8000

Table 22. Inhibitory action of formic acid against molds (Rehm 1961 b)

Name of test organism	pH value	minimum inhibitory concentration in ppm
Mucor racemosa	5.0	36000
Penicillium species	4.5–5.0	1640–80000
Gliocladium roseum	5.0	24000
Aspergillus species	5.0–7.0	15000–55000
Fungi imperfecti	5.0	12000–52000

16.10 Fields of Use

The importance of formates for use in preserving fish marinades has decreased greatly.

The same is true of pickled vegetables, for which calcium formate is still used on a small scale in combination with benzoates. A further reason for using calcium formate is that the calcium ions improve the firmness of the vegetable tissue and thus preserve the crisp texture of gherkins.

By adding 0.3–0.4% formic acid immediately after pressing, fruit juices and fruit pulps used in fruit juice production can be preserved against yeast fermentation. Formic acid has the advantage over other preservatives of being miscible with water. Consequently it can be incorporated in the juices very easily and readily gives a uniform mixture. In effective concentrations, formic acid is perceptible in both taste and odor, so its use is confined to fruit products intended for further processing. The use of formic acid as a preservative for fruit juices, fruit pulps and other fruit products has declined greatly in the last few years as it has been increasingly supplanted by other preservation methods.

16.11 Literature

Balatsouras GD, Polymenacos NG (1963) Chemical preservatives as inhibitors of yeast growth. J Food Sci 28, 267–275

Bremanis E (1949) Die photometrische Bestimmung des Formaldehyds mit Chromotropsäure. Z Anal Chem 130, 44–47

Jodin F-V (1865) Etudes sur quelques propriétes de l'acide formique. C Rend 61, 1179–1181

Life Sciences Research Office, Federation of American Societies for Experimental Biology (1976) Evaluation of the health aspects of formic acid, sodium formate and ethyl formate as food ingredients. PB-266282. National Technical Information Service. US Department of Commerce, Springfield

Lück H (1957) Katalasehemmung durch organische Säuren. Biochem Z 328, 411–419

Lück H (1958) Einfluß von Konservierungsmitteln auf Häminenzyme. I. Mitteilung. Allgemeine Wirkung. Z Lebensm Unters Forsch 108, 1–9

Lück H, Rickerl E (1969) Untersuchungen an *Escherichia coli* über eine Resistenzsteigerung gegen Konservierungsmittel und Antibiotica. Z Lebensm Unters Forsch 109, 322–329

Malorny G (1969a) Die akute und chronische Toxizität der Ameisensäure und ihrer Formiate. Z Ernährungswiss 9, 332–339
Malorny G (1969b) Stoffwechselversuche mit Natriumformiat und Ameisensäure beim Menschen. Z Ernährungswiss 9, 340–348
Palese M, Tephly TR (1975) Metabolism of formate in the rat. J Toxicol Environ Health 1, 13–24
Rehm H-J (1961a) Untersuchungen zur Wirkung von Konservierungsmittelkombinationen. V. Zur Frage der Resistenz von Mikroorganismen gegen Konservierungsmittel in Kombinaten. Z Lebensm Unters Forsch 115, 27–46
Rehm H-J (1961b) Grenzhemmkonzentrationen der zugelassenen Konservierungsmittel gegen Mikroorganismen. Z Lebensm Unters Forsch 115, 293–309
Reinefeld E, Bliesner K-M (1977) Enzymatische Bestimmung von Ameisensäure und ihre Anwendung auf Melassen. Zuckerind 30, 650–652
Scheer H (1971a) Über die konservierende Wirkung von Natriumformiat. Arch Fischereiwiss 22, 79–84
Scheer H (1971b) Über den Einfluß von Genuß- und Mineralsäuren auf die konservierende Wirkung von Natriumformiat. Arch Fischereiwiss 22, 263–279
Sporn A, Marin V, Schöbesch C (1962) Cercetări cu privire la toxicetatea acidului formic. Igiena 11, 507–515
Tanner H (1976) Photometrische Schnellbestimmung von Ameisensäure in Getränken. Schweiz. Z Obst Weinbau 112, 38–42
Tracor-Jitco, Inc (1974) Scientific literature reviews on Generally Recognized as Safe (GRAS) food ingredients – formic acid and derivates. PB-228558. National Technical Information Service, US Department of Commerce, Springfield

CHAPTER 17

Acetic Acid

17.1 Synonyms

IUPAC: Ethanoic acid
German: Essigsäure. *French:* Acide acétique. *Italian:* Acido acetico. *Spanish:* Acido acético. *Russian:* Уксусная кислота.

17.2 History

The pickling of foods in vinegar or vinegar-containing solutions ranks among the oldest methods of preservation; indeed, vinegar made from palm wine was known in the East as long ago as 5000 BC. In Ancient Rome, vinegar was used as a food additive for preservation purposes on its own or in admixture with salt, wine or honey (oxymel). Until the early Middle Ages vinegar was produced exclusively in the household, the method being to allow alcoholic liquids, especially wine, to acidify naturally through the influence of vinegar bacteria while exposed to air. Industrial production of vinegar commenced in France in the fourteenth century, using the Orléans process. Nowadays acetic acid continues to occupy a firm place in food preservation, and in some countries synthetic acetic acid has acquired considerable importance alongside vinegar produced by fermentation.

17.3 Commercially Available Forms, Derivatives

Acetic acid is used in food preservation in two forms, namely as 5 to 10 % vinegar and as 25 to 80 % aqueous solutions of synthetic acetic acid. The 5 to 10 % vinegar is obtained either by diluting synthetic acetic acid or blending acetic acid derived from fermentation and synthetic acetic acid, or by fermentation alone. Wine vinegar in Germany is often a blend of one part genuine wine vinegar and four parts spirit vinegar. Depending on the nature of the starting substance, a distinction is drawn between wine, fruit, beer, malt, spirit and other vinegars.

Sodium diacetate also is still of some importance as a preservative. It is a compound containing water of crystallization and comprising 1 mol each of acetic acid and sodium acetate.

17.4 Properties

CH_3COOH, molar mass 60.05, in pure form a colorless, transparent liquid solidifying at 17 °C, boiling at 118 °C and unrestrictedly miscible with water.

Sodium diacetate, white water-soluble powder with a crystalline structure and smelling of acetic acid.

17.5 Analysis

Acetic acid is volatile in steam. If no other volatile acids are present, it can be determined by titration of the steam distillate. Quantitative determination of acetic acid can be carried out by ion chromatography in conjunction with UV detection (Bouzas et al. 1991) or enzymatically.

17.6 Production

Acetic acid is manufactured by biological oxidation or by synthetic means.

All customary biological processes are based on the ability of acetobacter to oxidize ethanol to acetic acid. In former times surface techniques were in common use, the best-known of these being the Orléans, the Boerhaave, the generator and the Schützenbach or quick methods. Today, however, the submerge process is generally employed, the most important starting materials being wine and dilute alcohol (Conner and Allgeier 1976, Keszthelyi 1974, Greenshields 1978).

Acetic acid can be produced synthetically by oxidation of acetaldehyde or low hydrocarbons. Carbonylation of methanol is another method of some industrial importance.

17.7 Health Aspects

If they have identical acetic acid contents, vinegar obtained by fermentation and that derived from synthetic acetic acid do not differ in their toxicological properties (Bornmann et al. 1952).

17.7.1 Acute Toxicity

For dilute, non-caustic concentrations the LD_{50} of acetic acid for rats and mice after peroral administration is given as 3–5 g/kg body weight (von Oettingen 1960, Spector 1956). In concentrations above 30%, acetic acid will attack the skin. Because it is lipophilic, acetic acid has a more powerful denaturing effect than mineral acids with the same hydrogen ion concentration.

The maximum allowable workplace concentration for acetic acid is 25 mg/m^3.

17.7.2
Subchronic Toxicity

According to relatively old papers, rats tolerate drinking-water containing an addition of 0.01 to 0.25 % acetic acid over a period of 2 to 5 months without suffering any noticeable damage. This corresponds to a daily consumption of 0.2 g/kg body weight (Bornmann et al. 1952). Other investigations have revealed that rats tolerate 1.8 g dilute acetic acid daily over 2 weeks without damage, whereas a dose of 2.4 g daily will kill the animals within 3 to 5 days. Daily quantities of 4.2 to 4.8 g sodium acetate, however, are tolerated without reaction. It is concluded from the varying effects of acetic acid and acetates that the harmful action derives not from the acetate ion as such but from the continuous powerful action of the acid (von Oettingen 1960).

17.7.3
Chronic Toxicity

Only a few isolated reports deal with the chronic toxicity of acetic acid or acetates when used as food preservatives. The use of vinegar as a food ingredient for thousands of years is regarded as proving the harmlessness of acetic acid in low concentrations. Most of the studies into the toxicologically relevant properties and effects of acetic acid are concerned with health at work. Sensitization and intolerant reactions to acetic acid are very rare. The possibility that acetic acid may be directly carcinogenic or genotoxic can be ruled out.

Summary descriptions of the toxicology of acetic acid can be found in reports by BIBRA (BIBRA 1987) and the Federal German "MAK" committee (Hentschler 1993).

17.7.4
Biochemical Behavior

Acetic acid plays a central part in the metabolic process as acetyl co-enzyme A ("activated acetic acid"). It is involved in many biochemical reactions, i.e. the fatty acid and carbohydrate metabolism, and can be used as a source of energy by mammalian organisms.

17.8
Regulatory Status

As acetic acid occurs naturally in the human body and has also been used in food technology for thousands of years as a natural flavoring and acidulant, the addition of vinegar to foods is not subject to many food law restrictions in most countries. Some countries' regulations covering vinegar and acetic acid differentiate between vinegar obtained by fermentation and that manufactured synthetically. Hence, in a number of countries, especially those where agriculture and viticulture are practiced, although not in the Federal Republic of Germany, the term "vinegar" (or its equivalent in other languages) is confined for commercial reasons to vinegar obtained by fermentation. The European Union considers acetic acid (E 260) as a food additive.

17.9 Antimicrobial Action

So far as their preservative action is concerned, vinegar of synthetic origin and that manufactured by fermentation are identical, the only decisive factor being the concentration of the vinegar, i.e. the content of acetic acid.

17.9.1 General Criteria of Action

Even more than formic acid, the action of acetic acid is based essentially on lowering the pH value of the product to be preserved. Compared with other preservative acids, however, the concentrations of acetic acid required for this purpose are very large. Only above a concentration of some 0.5% does acetic acid display an antimicrobial action in penetrating the cell wall and denaturing the protein of the cell plasma (Reynaulds 1975).

If the food destined for preservation is adjusted to a pH of about 3 by the addition of acid, the antimicrobial effect of the acetic acid is 10 to 100 times as powerful as that of other acids, such as hydrochloric acid (Reynaulds 1975). This difference is due to the fact that undissociated acetic acid penetrates more readily to the interior of the cell because it is lipophilic. Acetic acid increases the heat-sensitivity of bacteria but not that of yeasts or molds.

Although acetic acid has an antimicrobial effect extending beyond its action on the pH, this is only slight. Its dissociation constant is comparable to that of propionic acid and sorbic acid; so much of it remains in undissociated form even in the high pH range, although, unlike sorbic acid and propionic acid, its effect is not great. Between pH 6 and pH 5 the action of acetic acid only doubles (Woolford 1975), whereas the undissociated portion increases about sevenfold over this range. There is therefore no positive correlation between the undissociated acid portion and antimicrobial efficacy. Consequently, acetates have no antimicrobial action of their own. The successful use of sodium diacetate does not contradict this statement, since its action is based on the free acetic acid it contains.

17.9.2 Spectrum of Action

The action of acetic acid is directed principally against bacteria, owing to its pH-reducing effect, since most bacteria strains have their growth optima in the weakly acid to neutral pH range. This is true most particularly of the pathogenic bacteria, including salmonella, although *Bacterium xylinum* is still perfectly viable in normal culinary vinegar. lactobacilli, too, are only slightly sensitive to acetic acid, since they have considerable acid tolerance (Woolford 1975). For this reason, amongst others, vinegar derived from fermentation is often treated with sulfur dioxide as protection against bacterial spoilage.

Although acetic acid is more effective against film-forming yeasts and molds than against bacteria, its action is still weak compared with other preservatives. At pH 5 the growth of common yeasts is retarded by additions of as little as 1% acetic acid. Growth is inhibited entirely in the presence of 3.5 to 4% acetic acid (Yamamoto et al. 1984b). Common salt improves the action of acetic acid, mainly by lowering the water activity (Yamamoto et al. 1984a). This effect has been used for a long time in vegetable and fish preservation. The effect of acetic acid against lactic acid bacteria is only slight (Yamamoto et al. 1984 a). Since acetic acid does not generally have a strong preserving action, it is often combined with physical methods of preservation, such as pasteurization, or with common salt and/or the more powerful preservatives sorbic acid or benzoic acid.

17.10 Fields of Use

17.10.1 Fat Products

Mayonnaises, salad creams and delicatessen salads all include varying quantities of acetic acid not only for reasons of flavor but also to improve their keeping properties. Nevertheless, the protection acetic acid affords against spoilage by bacteria, yeasts and molds is only limited; so in such cases it is often combined not only with salt but also with other methods of preservation, such as pasteurization or cold storage or other preservatives, such as sorbic acid and/or benzoic acid. Used in concentrations of over 1%, which would be necessary for preservation, acetic acid is unpleasant to the taste (Baumgart and Libuda 1977) but its excessive sharpness can be moderated with buffer substances such as sodium acetate (Stöltzing 1987).

17.10.2 Meat Products

The immersion or washing of fresh meat in dilute acetic acid is a food preservation measure practiced in the household.

17.10.3 Fish Products

Fish are frequently preserved in liquids containing vinegar and salt, prior to which they are tenderized by frying (fried marinades), boiling (boiled marinades) or cold processing (cold marinades), which involves treatment with salt and vinegar. Acetic acid is therefore of particular importance, especially since most preservatives in the narrow sense of the term are effective mainly against yeasts and molds, whereas the major factor in fish spoilage is bacterial activity. However, as the lactobacilli, which are especially important in this connection, are not reliably inhibited by acetic acid, even in concentrations of several percent, it is essential to combine the acetic acid with other preservation processes, e.g. pasteurization or cooling and/or other preservatives, in order to give marinades reliable keeping properties. The 1 to 3% concentrations of acetic acid which are usual for fish products provide basic protection against pathogenic microorganisms, especially clostridia.

17.10.4
Vegetable Products

The preservation of vegetables in vinegar corresponds to natural lactic acid fermentation although no fermentation process is involved. Raw vegetable products destined for preservation can be immersed in 0.5 to 3% acetic acid solutions, which may contain spices, salt and sugar or sweeteners, according to the product concerned. The main vegetable products preserved in acetic acid are gherkins, beetroots and mixed vegetables (mixed pickles). However, vinegar alone is insufficient to provide reliable long-term preservation, even for vegetable products, since film-forming yeasts and a number of molds can still develop well in vinegar preserving liquids containing as much as several percent acetic acid, especially if these liquids also contain sugar. For this reason acid-resistant sweeteners, such as acesulfame K, are often used as sugar substitutes. It is also advisable to pasteurize or sterilize the pickled vegetables. To prevent microbial contamination after containers are opened, especially those of relatively large size, sorbates and/or benzoates are added to the preserving liquids as well.

17.10.5
Fruit Products

In a number of countries plums, pears, cherries, grapes and other fruit are pickled in vinegar. The process involves covering the fruit with hot liquids containing 2 to 2.5% acetic acid and a large quantity of sugar, after which both fruit and liquids are sterilized.

17.10.6
Baked Goods

In the baked goods sector, acetic acid, preferably in the form of sodium diacetate, is used to protect a number of bread varieties against spoilage by *Bacillus mesentericus*, otherwise known as rope. The development of these bacteria can be retarded by slight acidification of the dough – a fact which has been known ever since 1906 (Watkins 1906). Acetic acid not only inhibits the bacteria themselves but also reduces their heat-resistance, so that their death rate during the baking process is higher in doughs acidified with acetic acid. Sodium diacetate is used in concentrations between 0.2 and 0.4%, relative to the quantity of flour. It has not been adopted for use in preventing mold on cut bread, however, as its antimicrobial action is inadequate.

17.11 Other Effects

Besides its preservative action, acetic acid is very important as a flavoring; indeed, in many foods its primary function is as a flavoring and its preservative function is of secondary importance. The action of acetic acid on protein may also have an influence on flavor. In low concentrations, acetic acid causes partial hydrolysis of the protein, especially in animal tissues, which may lead to the production of agreeably flavored cleavage products. This effect is important chiefly in the production of fish marinades (Meyer 1962). Whereas common salt tends to make fish meat firmer, vinegar has a tenderizing effect.

17.12 Literature

Baumgart J, Libuda H (1977) Haltbarkeit von Mayonnaisen und Feinkostsalaten in Abhängigkeit von Konservierungsstoff- und Essigsäureanteil. Z Lebensmittel-Technol Verfahrenstechn 28 (5), 181–182

BIBRA (1987) Toxicity profile acetic acid and its common salts

Bouzas J, Knatt C, Bodyfeldt F, Torres A (1991) Simultaneous determination of sugars and organic acids in cheddar cheese by high-performance liquid chromatography. J Food Sci 56, 276–279

Bornmann G, Küchle HJ, Loeser A, Meyer G, Stürmer E (1952) Über die Wirkung von Essig aus Essigessenz und Essig aus Äthylalkohol auf den Organismus. Z Lebensm Unters Forsch 94, 234–240

Conner HA, Allgeier RJ (1976) Vinegar: Its history and development. Adv Appl Microbiol 20, 81–133

Greenshields RN (1978) Acetic acid: Vinegar. Econ Microbiol 2, 121–186

Hentschler D (1993) Toxikologisch-arbeitsmedizinische Begründungen von MAK-Werten. Weinheim: VCH loose-leaf collection

Keszthelyi G (1974) Der heutige Stand der biologischen Herstellung von Essigsäure unter besonderer Berücksichtigung von submers arbeitenden Verfahren. Mitt Höheren Bundeslehr-Versuchsanst Wein Obstbau Klosterneuburg 24, 445–454

Meyer V (1962) Probleme des Verderbens von Fischkonserven in Dosen. VII. Untersuchungen über die Entstehung der Aminosäuren beim Marinieren von Heringen. Veröff Inst Meeresforsch Bremerhaven 8, 21–36

Oettingen WF von (1960) The aliphatic acids and their esters – toxicity and potential dangers. The saturated monobasic aliphatic acids and their esters. Acetic acid and esters: Acetic acid. Amer Med Assoc Arch Ind Health 21, 28–32

Reynolds AE (1975) The mode of action of acetic acid on bacteria. Diss Abstr B, 35, 4935–4936

Spector WS (1956) Handbook of toxicology, Vol 1. WB Saunders, Philadelphia, pp. 44–45 and 56–57

Stöltzing U (1987) Einfluβ von Essigsäure und Puffersubstanzen auf die Haltbarkeit und die sensorischen Eigenschaften unkonservierter Salatmayonnaise. Lebensmitteltechn 19 (5), 96–99

Watkins EJ (1906) Ropiness in flour and bread and its detection and prevention. J Soc Chem Ind (London) 25, 350–357

Woolford MK (1975) Microbiological screening of the straight chain fatty acids (C_1 – C_{12}) as potential silage additives. J Sci Food Agric 26, 219 – 228
Yamamoto Y, Higashi K, Yoshii H (1984 a) Inhibitory activity of organic acids on food spoilage bacteria. (Studies on growth inhibition of food spoilage microorganisms for low salt foods. Part I). Nippon Shokuhin Kogyo Gakkaishi 31, 525 – 530
Yamamoto Y, Higashi K, Yoshii H (1984 b) Inhibitory activity of acetic acid on yeasts (Studies on growth inhibition of food spoilage microorganisms for low salt foods. Part III) Nippon Shokuhin Kogyo Gakkaishi 31, 772 – 776

CHAPTER 18

Propionic Acid

18.1 Synonyms

IUPAC: Propanoic acid
German: Propionsäure. *French:* Acid propionique. *Italian:* Acido propionico. *Spanish:* Acido propiónico. *Russian:* Пропионовая кислота.

18.2 History

The fact that propionic acid and its salts have an antimicrobial action has been known for a long time. Its planned use in the preservation of baked goods was first proposed in 1938 (Hoffman et al.), although the efficacy of organic acids against rope in baked goods had been recognized long before (Watkins 1906). Since the end of the nineteen-thirties propionates have been used in the USA on a large scale in the preservation of bread and on a smaller scale to preserve cheese. Propionates have also become well established for baked goods preservation in other countries, where they are used principally for low-acid white bread.

18.3 Commercially Available Forms, Derivatives

The main forms used in food technology are sodium and calcium propionate. Only in a few countries is pure propionic acid important as a preservative for human foods, its main field of use being in animal feed.

18.4 Properties

CH_3CH_2COOH, molar mass 74.08, in pure form a colorless, transparent liquid with a disagreeable, pungent odor, unlimited miscibility with water and a boiling point of 141 °C. Sodium propionate, molar mass 96.06 and calcium propionate, molar mass 186.22, white, slightly water-soluble powders with odors resembling propionic acid.

18.5 Analysis

Like many other preservatives, propionic acid can be identified by IR spectroscopy (Losada 1989).

Propionic acid is readily volatile in steam. Hence, like its salts, it can be detected by paper or thin-layer chromatography in steam distillate. No specific color reactions are known. Quantitative determination by gas chromatography is possible (Lück et al. 1975). A more precise method of quantitative determination is described in the Swiss Foodstuffs Manual. After derivatization to *p*-nitrobenzyl ester, propionic acid can be determined by high performance liquid chromatography with subsequent UV detection at 265 nm (Badoud and Pratz 1986).

18.6 Production

Propionic acid is a by-product from the oxidation of hydrocarbons by air in acetic acid production. It can be obtained directly by the conversion of ethylene with CO and water, using the Reppe method. The propionates are formed by the neutralization of propionic acid with the corresponding metal hydroxides.

18.7 Health Aspects

18.7.1 Acute Toxicity

The LD_{50} of propionic acid (rat, oral) is 2.6 g/kg body weight (BIBRA 1987). The acute toxicity of sodium and calcium propionate is in the same order of magnitude. The acute toxicity of calcium propionate is not increased by combination with other preservatives (Sado 1973). In concentrated form propionic acid irritates the skin and mucous membranes.

The maximum allowable workplace concentration for propionic acid is 30 mg/m^3.

18.7.2 Subchronic Toxicity

Young rats tolerate a diet containing 1 to 3% sodium or calcium propionate over several weeks and develop just as well as control animals (Harshbarger 1942). The administration of a diet containing 24% propionate to young rats kills them after 5 days, whereas adult rats tolerate such a diet for some 20 days despite sustaining damage (McAtee et al. 1968). The intestinal flora of pigs is not influenced by a diet containing 0.8% propionic acid over 4 weeks (Cole et al. 1968). Neither propionic acid nor calcium propionate have a mutagenic effect (Litton Bionetics 1976).

18.7.3
Chronic Toxicity

After rats were exposed to 4% propionic acid in the feed over 20 weeks, hyperplasia and papillomas in the mucous membrane of the forestomach were observed (Griem 1985). Histological investigations during these experiments revealed basal cell growths and carcinomatose changes in the plate epithelium in the mucous membrane of the forestomach. Hyperplastic changes were also detected in the group of experimental animals whose feed contained 0.4% propionic acid. In subsequent experiments it was demonstrated that propionic acid is not genotoxic and that there is no genetically related mechanism for inducing plate epithelium tumors in the rat forestomach. In subchronic experiments carried out on other animal species (e.g. Beagle) with 0.3–3% propionic acid in the feed, no changes in the mucous membrane of the stomach and esophagus were detected. Studies (Harrison et al. 1991, Boeld and Netter 1993) indicate that other short-chain aliphatic carboxylic acids (e.g. butyric acid) can also induce changes in the rat forestomach similar to those described for propionic acid. These studies also show that the composition of the feed has a strong influence on the induction of tumors in the rat forestomach by propionic acid.

In experiments with propionic acid on various species of rodent to determine any toxicological effects on the reproductive mechanism, no unfavorable effects induced by propionic acid were detected (BIBRA 1987, Harrison 1992).

According to a statement by the SCF, it is assumed that effects on human health due to the use of propionic acid as a food additive in the quantities currently employed are unlikely. However, the results of further comparative studies with related short-chain fatty acids and their salts are still awaited (SCF 1990).

18.7.4
Biochemical Behavior

Propionic acid and propionates are readily absorbed by the digestive tract on account of their good water-solubility. There is no risk of their accumulation in the body, as propionic acid is metabolized in the body in a manner similar to that of fatty acids (Elliot et al. 1965, Bässler 1959). Some of the propionic acid is incorporated into glucose, glycogen and other metabolic products. Even after the administration of large doses in the diet, none is excreted in the urine. The decomposition of propionic acid in the tissue of mammals takes place by linkage with coenzyme A via methylmalonyl-CoA, succinyl-CoA and succinate to give CO_2 and H_2O (Bässler 1959). Besides the mode of decomposition described above, it is presumed that propionic acid can also decompose by way of succinate to give β-alanin (Rendina and Coon 1957). Among ruminants propionate also condenses directly with acetate to form higher fatty acids with odd numbers of carbon atoms (Bässler 1959). Since propionic acid is formed by the decomposition of a number of amino acids and by the oxidation of fatty acids containing odd numbers of carbon atoms, it is not a substance foreign to the body but a physiological intermediate product of the normal metabolism (Bässler 1959).

18.8
Regulatory Status

Virtually all countries with industrialized bread production permit the use of sodium and calcium propionate, and some countries even allow propionic acid itself, for the preservation of bread and in certain instances also other baked goods.

18.9
Antimicrobial Action

18.9.1
General Criteria of Action

A number of microorganisms can produce propionic acid, while many others can metabolize existing propionic acid. Should propionic acid be present in relatively large concentrations, as is the case with food preservatives, its inhibitory action is due largely to the fact the it accumulates in the cell and blocks metabolism by inhibiting enzymes. Depending on the concentration, propionic acid also lowers the intracellular pH value. This effect also helps to inhibit growth and kill the cells (Salmond et al. 1984). It is more pronounced in the undissociated portion of the acid than in the dissociated portion (Eklund 1985).

With propionic acid, like other preservatives based on carboxylic acids, the pH value of the substance to be preserved is of great importance to the antimicrobial action. Owing to its low dissociation constant, propionic acid behaves in this respect in a favorable manner resembling that of sorbic acid and can therefore be used for preserving foods with a high pH value.

The antimicrobial action of propionic acid is weak in comparison with other preservatives. In practical food preservation, propionic acid therefore has to be employed in relatively high dosages.

18.9.2
Spectrum of Action

Owing to the unspecific mechanism of action of propionic acid, its spectrum of action cannot be delineated precisely. Practical experience has shown that the action of propionic acid and propionates is directed primarily against molds (von Schelhorn 1951), although some strains of Penicillium exist which even grow on nutrient media containing over 5% propionic acid (Heseltine 1952). Yeasts are likewise inhibited, and so also are bacteria, although principally gram-negative strains. A number of yeasts, e.g. *Torula* varieties, are capable of utilizing propionic acid in their metabolism. A factor of practical importance is the efficacy of the propionic acid and propionates against *Bacillus mesentericus*, the bacterium that causes rope in bread.

18.10
Fields of Use

18.10.1
Dairy Produce

Propionic acid bacteria play a major part in the ripening of several cheese varieties, e.g. Emmentaler. Such cheeses therefore contain their own "natural" propionic acid although not in concentrations capable of preventing undesired mold formation. In the USA 5 to 10 % solutions of sodium or calcium propionate are used on a small scale for the surface treatment of hard cheese against mold attack. Processed cheese can be protected against mold attack by the addition of 0.2 to 0.3 % propionate. The use of propionic acid in the cheese sector, which has in any case never been of great significance, is now negligible.

18.10.2
Baked Goods

In all countries where baked goods are manufactured on an industrial scale, propionates are extremely important as preservatives for bread and cakes of all kinds. The low dissociation constant of propionic acid means they remain very effective in the high pH range in which most baked goods normally fall and they effectively combat molds and *Bacillius mesentericus*, the bacterium that causes rope. Nevertheless, the low absolute efficacy of the propionates means that relatively large concentrations are needed in order to keep bread and other baked goods free of mold for more than a few days. Such concentrations inhibit leavening, and this effect must be offset by increasing yeast additions or extending the proving time of the dough. Another consideration is that propionates in the quantities necessary for a reasonable preservative effect give the bread a distinct odor, which is most marked in sliced bread when it is toasted. The propionates used are exclusively those of sodium and calcium, the former being employed chiefly in flour confectionery and the latter mainly for bread. For organoleptic reasons little, if any, use is made of pure propionic acid for bread preservation.

Propionates are added during the dough production stage, the concentration depending on the nature of the product and desired storage life of the baked goods in question. The quantities normally added are 0.1 to 0.3 %, relative to the weight of the flour.

Propionates are used in baked goods not only for economic reasons but also on account of their action against mycotoxin-forming molds (Reiß 1976).

It used to be believed that appreciable quantities of propionic acid formed in the dough during leavening; so attempts were made to breed cultures of propionic acid bacteria which would provide bread with its own protection against mold (Seiler 1964). However, it has now been proved that, at least in Germany, propionic acid does not occur naturally in sour dough or in bread raised with it (Lück et al. 1975, Rabe 1978).

18.11 Other Effects

In high dosages, propionic acid and its salts may impart a distinct flavor and odor to foods they are used to preserve. In a number of countries, e.g. the US, propionic acid is permitted and employed as a constituent of flavors.

18.12 Literature

Badoud R, Pratz G (1986) Improved high performance liquid chromatographic analysis of some carboxylic acids in food and beverages as their *p*-nitrobenzyl esters. J Chromatography 360, 119–136

Bässler KH (1959) Stoffwechsel und Stoffwechselwirkungen von Propionsäure im Hinblick auf ihre Verwendung als Konservierungsmittel. Z Lebensm Unters Forsch 110, 28–42

BIBRA, British Industrial Biological Research Association (1987) Toxicity profile, Propionic acid and its common salts. Carshalton, UK

Boeld J, Netter K (1993) Factors affecting distribution of ingested propionic acid in the rat forestomach. Food Chem Toxic 31, 169–176

Cole DJA, Beal RM, Luscombe JR (1968) The effect on performance and bacterial flora of lactic acid, propionic acid, calcium propionate and calcium acrylate in the drinking water of weaned pigs. Vet Rec 83, 459–464

Eklund T (1985) Inhibition of microbial growth at different pH levels by benzoic and propionic acids and esters of *p*-hydroxybenzoic acid. Int J Food Microbiol 2, 159–167

Elliot JM, Hogue DE, Myers GS, Loosli JK (1965) Effect of acetate and propionate on the utilization of energy by growing-fattening lambs. J Nutr 87, 233–238

Griem W (1985) Tumorogene Wirkung von Propionsäure an der Vormagenschleimhaut von Ratten im Fütterungsversuch. Bundesgesundheitsbl 28, 322–327

Harrison P, Grasso P, Badescu V (1991) Early changes in the forestomach of rats, mice and hamsters exposed to dietary propionic and butyric acid. Food Chem Toxicol 29, 367–371

Harrison P (1992) Propionic acid and a phenomenon of rodent forestomach tumorigenesis: A review. Food Chem Toxicol 30, 333–340

Harshbarger KE (1942) Report of a study on the toxicity of several food-preserving agents. J Dairy Sci 25, 168–174

Heseltine WW (1952) Sodium propionate and its derivatives as bacteriostatics and fungistatics. J Pharm Pharmcol 4, 577–581

Hoffman C, Dalby G, Schweitzer TR (1938) (Ward Baking Company, New York): Process for inhibition of mold. US Patent 2 154–449

Litton Bionetics, Inc (1976) Mutagenic evaluation of compound FDA 75–62 Propionic acid. PB-266 897. National Technical Information Service. US Department of Commerce, Springfield; and Mutagenic evaluation of compound FDA 75–61 Sodium propionate NF, FCC. PB-266 900. National Technical Information Service. US Department of Commerce, Springfield

Losada P, Lozano J, Gandara J (1989) Contribución a la identificación de aditivos conservadores. Alimentaria 204, 57–83

Lück E, Oeser H, Remmert K-H, Sabel J (1975) Über das Nichtvorkommen von Propionsäure in Roggen- und Mischbrot. Z Lebensm Unters Forsch 158, 27–29

McAtee JW, Little CO, Mitchell GE (1968) Utilization of rumen metabolites as energy sources in rats. Life Sci 7 II, 769–775

Rabe E (1978) Propionsäure in Sauerteig und Brot. Getreide, Mehl, Brot 32, 85–89

Reiß J (1976) Mycotoxine in Lebensmitteln. VIII. Hemmung des Schimmelpilzwachstums und der Bildung von Mycotoxinen (Aflatoxine B_1 and G_1, Patulin, Sterigmatocystin) in Weizenvollkornbrot durch Sorbinsäure und Sorboylpalmitat. Dtsch Lebensm Rundsch 72, 51–54

Rendina G, Coon MJ (1957) Enzymatic hydrolysis of the coenzyme A, thiol esters of β-hydroxypropionic and β-hydroxyisobutyric acids. J Biol Chem 225, 523–534
Sado I (1973) Synergistic toxicity of official permissible preservative food additives. Nippon Eiseigaku Zasshi 28, 463–476
Salmond CV, Kroll RG, Booth IR (1984) The effect of food preservatives on pH homeostasis in *Escherichia coli.* J Gen Microbiol 130, 2845–2850
Schelhorn M von (1951) Wirksamkeit und Wirkungsbereich chemischer Konservierungsmittel für Lebensmittel. Z Lebensm Unters Forsch 92, 256–266 and Untersuchungen über Konservierungsmittel. V. Zur Frage des Vergleichs der Wirksamkeit von Konservierungsmitteln. Dtsch Lebensm Rundsch 47, 16–18
Schweizer Lebensmittelbuch (Swiss Foodstuffs Manual), new edition: Konservierungsmittel für Lebensmittel, chapter 44, no 7, "Bestimmung der Propionsäure und ihrer Salze in Backwaren", Bern, 1st instalment 1992
Seiler DAL (1964) Factors affecting the use of mould inhibitors in bread and cake. Proc 4th Int Symp Food Microbiol Göteborg, p. 221–220
Watkins EJ (1906) Ropiness in flour and bread and its detection and prevention. J Soc Chem Ind (London) 25, 350–357
Wissenschaftlicher Lebensmittelauschuß der Europäischen Gemeinschaft (SCF) (1990) Bericht des Wissenschaftlichen Lebensmittelausschusses dated October 19, 1990, page 11, "Propionsäure"

Chapter 19

Sorbic Acid

19.1 Synonyms

IUPAC: Hexa-2,4-dienoic acid
English: 2,4-hexadienoic acid. *German:* Sorbinsäure, *trans-trans*-2,4-Hexadiensäure (1). *French:* Acide sorbique. *Italian:* Acido sorbico. *Spanish:* Acido sórbico. *Russian:* Сорбиновая кислота.

19.2 History

The antimicrobial action of sorbic acid, first manufactured by Hofmann from rowan berry oil in 1859, was discovered in Germany by Müller in 1939 and independently, a few months later, by Gooding in the USA. Sorbic acid first became available from industrial production in the mid-1950s and has since been used to a growing extent for food preservation throughout the world. It is now increasingly preferred to other preservatives because of its physiological harmlessness and organoleptic neutrality.

19.3 Commercially Available Forms, Derivatives

Sorbic acid is used both as a free acid and as its potassium and calcium salts in various forms (powder, granules, solutions). The esters of sorbic acid with low aliphatic alcohols, which likewise have a preservative action, are of no importance as food preservatives, owing to their powerful self-odor.

19.4 Properties

CH_3–CH=CH-CH=CH–COOH, molar mass 112.13, white monoclinic crystals with a faint specific odor and sour taste which melt at 132 to 135 °C. At room temperature 0.16 g sorbic acid dissolve in 100 g water and 0.07 g in 100 g 10% sodium chloride solution. Some 13 g sorbic acid dissolve in 100 g anhydrous ethanol or in 100 g glacial acetic acid. Its solubility in fatty oils is 0.5 to 1 g per 100 g, depending on the type of oil in question.

Potassium sorbate, molar mass 150.22, white powder or granules. The most readily soluble of the sorbates. At room temperature, 138 g potassium sorbate dissolves in 100 g water. Up to 54 g potassium sorbate dissolves in 100 g 10% sodium chloride solution.

Calcium sorbate, white, odorless and tasteless powder resembling talcum. Solubility in water 1.2 g/100 g.

In solid form, sorbic acid, potassium sorbate and especially calcium sorbate are very stable, despite the two double bonds in the molecule. In solutions, the presence of oxygen causes oxidative degradation which may result in brown discoloration (Thakur et al. 1994). In commercial food preservation this is of no importance, since treated foods are generally consumed before any appreciable degradation occurs. Many other food ingredients, e.g. fat and flavorings, are in any case much more sensitive to oxidation than sorbic acid.

19.5 Analysis

Because of its volatility in steam, sorbic acid can be quantitatively isolated by acid steam distillation from the food to be investigated. The criterion used for qualitative detection and quantitative determination is the red coloration that sorbic acid produces with 2-thiobarbituric acid after oxidation with potassium dichromate (Schmidt 1960). As a polyunsaturated compound, sorbic acid displays a pronounced absorption maximum at some 260 nm (depending on the pH of the solution), which can be likewise used for quantitative determination (Luckmann and Melnick 1955). HPLC has emerged as the preferred method for determining sorbic acid and is sometimes used in multimethods, which can be employed for detecting benzoic acid, salicylic acid, parabens and sorbic acid simultaneously. In most cases RP-18 phases are used as stationary phases, with UV detection at wavelengths of 230 nm. Methods have been published both for detecting sorbic acid in foods in general (Bui and Kooper 1987, Hagenauer-Hener et al. 1990, Reifschneider et al. 1994) and for detecting sorbates and other preservatives specifically in cheese (Küppers 1988), yogurt (Olea Serrano et al. 1991), fruit juices (Kantasubrata and Imamkhasani 1991) and wine (Flak and Schaber 1988).

Standardized methods of detecting sorbates (GC, TLC and HPLC) have been published in the revised edition of the Swiss Foodstuffs Manual (1992). There is also a method for detecting them in liquid tabletop sweeteners in accordance with § 35 of the Federal German food law (L.57.22.99). Rather unconventional techniques of detecting sorbic acid by ion chromatography or capillary isotachophoresis (Karovicova et al. 1991) have not so far become established in routine use. X-ray structural analysis of sorbic acid has also been described (Cox 1994).

19.6 Production

Nowadays the only industrial production method used for sorbic acid is that employing ketene and crotonaldehyde. A polymeric ester forms as an intermediate (Lück 1993). The production of sorbic acid by the oxidation of 2,4-hexadienal has ceased to be of any significance.

19.7 Health Aspects

19.7.1 Acute Toxicity

The LD_{50} of sorbic acid (rat, oral) is 10.5 g/kg body weight (Deuel et al. 1954). Other authors quote LD_{50} values of 7.4 g/kg (Smyth and Carpenter 1948) and 8.7 g/kg body weight (Sado 1973). The acute toxicity of sorbic acid is not changed by feeding the experimental animals with other preservatives at the same time (Sado 1973).

Sorbic acid irritates the mucous membranes, but only in highly sensitive persons does it cause irritation to intact skins. The allergenic potential of sorbic acid is assessed as being extremely low, since, as a low-molecular substance, it cannot cause an antibody response. Moreover, covalent bonds with proteins, which might result in immediate hypersensitivity, are not known (Vieths et al. 1994). Pseudo-allergic reactions to sorbic acid as a food additive are relatively rare (Rosenhall 1982, Hannuksela and Haahtela 1987, Häberle 1989). Its acute fish toxicity of 1250 to 1800 mg/liter is extremely low so that it qualifies for inclusion in the lowest Federal German water hazard class (Koch 1994). Sorbic acid is readily biodegradable in the soil and wastewater (Lück 1993).

19.7.2 Subchronic Toxicity

10% sorbic acid in the feed was tolerated by rats for 42 days without damage (Cremer et al. 1959). In another feeding experiment an equally high dosage of sorbic acid added to the feed of rats over a period of 120 days led to increased growth and greater liver weight. The reproductive behavior of the animals remained normal (Demaree et al. 1955). Additions of 5% sorbic acid to the feed of rats and dogs produced no damage whatever during a 90-day experiment; even 8% sorbic acid, representing 5 g/kg body weight, caused only a slight increase in liver weight without histological changes (Deuel et al. 1954 a). A number of further investigations largely confirm these findings (Food and Drug Research Laboratories 1973, BIBRA 1987, Walker 1990). The increase in liver weight occasionally observed is interpreted as functional hypertrophy, the greater weight increase being attributable to the caloric utilization of the sorbic acid (Lang 1960).

19.7.3 Chronic Toxicity

Sorbic acid was not introduced for food preservation until the time when protracted tests of chronic toxicity for new food additives were being increasingly demanded. Consequently, sorbic acid is probably the most thoroughly investigated of all preservatives.

Throughout their entire lives a group of rats received an addition of 5% sorbic acid in their feed without any sign of damage. All bodily functions investigated remained normal, even in the following generation. Among the male rats there was even evidence of a temporary, statistically significant acceleration in growth and an increase in life span (Lang 1960). Sorbic acid fed to rats and mice in quantities of 40 to 80 mg per kg body weight daily over 17 to 18 months produced no untoward effects (Shtenberg and Ignat'ev 1970). In another test continued over two years, the feed administered to rats contained 1.5% and, in a further case, 10% sorbic acid. Among rats fed with the lower concentration no abnormalities could be observed in comparison with the control animals so far as growth, blood picture, or the condition and function of 12 internal organs were concerned. The addition of 10% sorbic acid to the feed led to a slightly reduced weight increase and also to enlargement of the thyroid gland, liver and kidneys (Gaunt et al. 1975). Similar results were obtained with mice (Hendy et al. 1976).

Sorbic acid administered in the feed has no carcinogenic action (Dickens et al. 1968, Shtenberg and Ignat'ev 1970, Gaunt et al. 1975, Hendy et al. 1976); neither has potassium sorbate (Dickens et al. 1968, Walker 1990).

Extensive mutagenicity studies have been conducted with potassium sorbate, all of which produced negative results (Litton Bionetics 1974). The results of similar tests with calcium sorbate were likewise negative (Litton Bionetics 1977). Potassium sorbate has also been shown to be non-teratogenic (Food and Drug Research Laboratories 1975).

In certain short-duration in-vitro genotoxicity tests, sodium sorbate stored in the air and sodium sorbate solutions were shown to have weak genotoxic effects (Münzner et al. 1990, Schiffmann and Schlatter 1992, Schlatter et al. 1992, Würgler et al. 1992). Sodium sorbate is known to be unstable both in solution and especially in solid form. For this reason it is not produced commercially. An oxidation product of sodium sorbate, 4,5-epoxy-2-hexenoic acid, has been found to be responsible for its genotoxic effect. Non-oxidized sodium sorbate solution, potassium sorbate and sorbic acid stored in the absence of air do not form this epoxide and display no genotoxic potential, either in vitro or in vivo (Jung et al. 1992, Schlatter et al. 1992, Würgler et al. 1992).

19.7.4 Biochemical Behavior

Earlier in-vitro investigations suggested that sorbic acid is metabolized like other fatty acids. This releases 27.6 kJ/g, of which some 50% is biologically utilizable (Lück 1972). In the human and animal organism, sorbic acid is subject to b-oxidation typical of fatty acid degradation (Table 23) (Deuel et al. 1954 b, Lang 1960).

In the case of extremely high dosages there is also evidence of slight ω-oxidation analogous to that occurring in conjunction with the usual nutrient fatty acids (Kuhn et al. 1937).

After 61 to 1213 mg 1-^{14}C-sorbic acid per kg body weight is administered to rats in their feed, 85% of the sorbic acid is expired as $^{14}CO_2$, irrespective of the dosage. The half-life, which depends on the dosage, is between 40 and 110 minutes. No sorbic acid is excreted in the urine. Some of the sorbic acid is used in the synthesis of new fatty acids by way of the acetyl-CoA occurring in the decomposition, since about 13% of the radioactivity is found in the internal organs, musculature

Table 23. Metabolic degradation of sorbic acid in the human and animal organism

CH_3-CH=CH-CH=CH-COOH	+ H_2O	hydration by enoylhydratase
CH_3-CH=CH-CHOH-CH_2-COOH	- 2H	dehydrogenation by β-ketohydrase
CH_3-CH=CH-CO-CH_2-COOH	+ $2O_2$	oxidation by β-ketothiolase
CH_3-CH=CH-COOH + $2CO_2$ + H_2O	+ H_2O	hydration by enoylhydratase
CH_3-CHOH-CH_2-COOH	- 2H	dehydrogenation by β-ketohydrase
CH_3-CO-CH_2-COOH	+ $2O_2$	oxidation by β-ketothiolase
CH_3-COOH + $2CO_2$ + H_2O		

and carcass of the animals (Fingerhut et al. 1962). These findings were confirmed elsewhere on mice (Westöö 1964).

19.8 Regulatory Status

Sorbic acid, potassium and calcium sorbate are permitted in all countries of the world for the preservation of many foods. The maximum permissible quantities, other than in exceptional cases, are between 0.1 and 0.2%. There is currently a recognizable worldwide trend towards using sorbic acid in place of other less well-tested preservatives on account of its harmlessness to health.

19.9 Antimicrobial Action

19.9.1 General Criteria of Action

The antimicrobial action of sorbic acid is based on several factors. Its inhibition of various enzymes in the microorganism cell will be discussed first. Enzymes of the carbohydrate metabolism such as enolase (Azukas 1962) and lactate dehydrogenase (Rehm 1967) are particularly important in this regard. Sorbic acid also intervenes relatively powerfully, though not very specifically, in the citric acid cycle, inhibiting malate dehydrogenase, isocitrate dehydrogenase, α-ketoglutarate dehydrogenase (Rehm 1967), succinate dehydrogenase (Rehm and Wallnöfer 1964), fumarase and aspartase (York and Vaughn 1964) among others. In addition, sorbic acid forms covalent bonds with SH groups of enzymes by way of its own double bonds, the SH groups thereby being inactivated (Martoadiprawito and Whitaker 1963). Furthermore, because sorbic acid is known to act against catalase-positive microorganisms, its influence on catalase and peroxidase is interesting (Lück 1957, Lück 1958). The inhibitory effect of sorbic acid is unlikely to be due to inhibition of a single enzyme. The points of attack in the cell may well differ in bacteria, yeasts and molds (Liewen and Marth 1985).

Another target for the action of sorbic acid being discussed is the cell wall (Eklund 1980, Eklund 1981, Eklund 1985). Sorbic acid inhibits, for example, the growth of *Escherichia coli* and the absorption of amino acids, such as serine and alanine, even in lower concentrations than the nucleic acid synthesis and the activity of lactate dehydrogenase (Eklund 1981). Partial destruction of the cell membrane increases the flow of protons into the cell, which has to use a larger amount of energy to counteract the potential differences that occur (Eklund 1985).

For sorbic acid to be able to develop its action within the microorganism cell, it must penetrate the cell wall. When this happens, it is primarily the undissociated acid constituent that enters the cell. Thus, at pH 3.15 some 40 % of the available sorbic acid enters the cell interior, whereas at the neutral point 99 % of the sorbic acid remains in the substrate (Oka 1960). This fact explains how the sorbic acid's action depends on the pH. For food preservation the undissociated proportion of the sorbic acid has by far the most effective action. Owing to its low dissociation constant of $1.73\ 10^{-5}$, sorbic acid, unlike other preservative acids, can also be employed for preserving weakly acid foodstuffs with a high pH. There are reasons to believe that sorbic acid has an antimicrobial action even in the form of the dissociated molecule. However this action is about 100 times weaker than that of the undissociated acid (Rehm and Lukas 1963, Eklund 1983). Neither in the case of *Escherichia coli* (Lück and Rickerl 1959) nor that of fungi (Lukas 1964, Viñas et al. 1990) does resistance in the true sense of the term occur, i.e. a rise in the minimum inhibitory concentration under the influence of sublethal sorbic acid concentrations. Industrial experience over 40 years with sorbic acid in food preservation confirms this fact.

Some microorganisms are capable of metabolizing sorbic acid if it is present in sublethal concentrations and the microbes counts are high. Consequently, sorbic acid is not suitable for the "preservation" of substrates with a high microbe count; so its use is restricted to preserving food with low microbe counts and of excellent hygienic standard. Extremely thorough investigations have been conducted into the decomposition of sorbic acid by aspergillus (Lukas 1964) and penicillium (Finol et al. 1982, Kinderlerer and Hatton 1990).

In the alkaline pH range the formation of methyl ketones was observed (Lukas 1964). *Penicillium roqueforti* is able to form 1,3-pentadiene (Marth et al. 1966), which has a highly unpleasant odor. A few strains of lactic acid bacteria are capable of reducing sorbic acid to the corresponding alcohol, hexadienol (Radler 1976), which can react chemically with ethanol to give 1-ethoxy-2,4-hexadiene and 2-ethoxy-3,5-hexadiene (Crowell and Guymon 1975, von Rymon Lipinski et al. 1975).

19.9.2 Spectrum of Action

By and large, the action of sorbic acid is directed primarily against yeasts and molds, including aflatoxin-forming microorganisms (Wallhäuber and Lück 1970). Among bacteria, the catalase-positive are inhibited more than the catalase-negative (Emard and Vaughn 1952), the strictly aerobic bacteria the most (York 1950), and both lactic acid bacteria and clostridia the least (York and Vaughn 1955). Assertions in the literature that sorbic acid has virtually no action against clostridia are based on nutrient medium tests conducted at pH 7, which is the optimum value for these microorganisms. Because of dissociation factors, sorbic acid has virtually no effect in this range in any case. Other investigations on meat products have revealed that sorbic acid very probably also inhibits clostridia as well as their formation of toxins if it is combined with small, intrinsically ineffective concentrations of nitrite and/or common salt and/or phosphates and the pH value is slightly reduced (Ivey and Robach 1978, Ivey et al. 1978, Leistner et al. 1978).

The effect of sorbic acid on various strains of microorganisms has been investigated in a number of general studies (Sofos and Busta 1981, Liewen and Marth 1985, Sofos 1992). Minimum inhibitory concentrations of sorbic acid when used against some of the bacteria, yeasts and molds involved in food spoilage are given in Tables 24 to 26 (Rehm 1961, Lück 1972). Most of the minimum inhibitory concentrations given in the Tables are based on nutrient medium tests. As the substrate is influenced by many factors (see Sect. 5.7), these concentrations serve merely as a guide for industrial and commercial food preservation.

Table 24. Inhibitory action of sorbic acid on bacteria

Name of test organism	pH value	Minimum inhibitory concentration in ppm
Pseudomonas species	6.0	1000
Micrococcus species	5.5–6.4	500– 1500
Pediococcus cerevisiae		1000
Lactobacillus species	4.4–6.0	2000– 7000
Achromobacter species	4.3–6.4	100– 1000
Escherichia coli	5.2–5.6	500– 1000
Serratia marcescens	6.4	500
Bacillus species	5.5–6.3	500–10000
Clostridium species	6.7–6.8	> 1000
Salmonella species	5.0–5.3	500–10000

Table 25. Inhibitory action of sorbic acid on yeasts

Name of test organism	pH value	Minimum inhibitory concentration in ppm
Saccharomyces cerevisiae	3.0	250
Saccharomyces ellipsoideus	3.5	500–2000
Saccharomyces species	3.2–5.7	300–1 000
Hansenula anomala	5.0	5000
Brettanomyces versatilis	4.6	2000
Byssochlamys fulva	3.5	500–2500
Rhodotorula species	4.0–5.0	1000–2000
Torulopsis holmii	4.6	4000
Torula lipolytica	5.0	1000–2000
Kloeckera apiculata	3.5–4.0	1000–2000
Candida krusei	3.4	1000
Candida lipolytica	5.0	1000

Table 26. Inhibitory action of sorbic acid on molds

Name of test organism	pH value	Minimum inhibitory concentration in ppm
Rhizopus species	3.6	1200
Mucor species	3.0	100–1000
Geotrichum candidum	4.8	10000
Oospora lactis	2.5–4.5	250–2000
Trichophyton mentagrophytes		1000
Penicillium species	3.5–5.7	200–1000
Penicillium digitatum	4.0	2000
Penicillium glaucum	3.0	1000–2500
Aspergillus species	3.3–5.7	200–1000
Aspergillus flavus		1000
Aspergillus niger	2.5–4.0	1000–5000
Botrytis cinerea	3.6	1200–2500
Fusarium species	3.0	1000
Cladosporium species	5.0–7.0	1000–3000

19.10 Fields of Use

19.10.1 Fat Products

In comparison with other preservatives, sorbic acid has a favorable oil/water distribution coefficient; so in fat/water emulsions a relatively high proportion of sorbic acid/sorbate remains in the water phase, which is the only one to be microbiologically susceptible. Hence, concentrations of 0.05 to 0.1% sorbic acid are used for the preservation of margarine (Becker and Roeder 1957). Sorbic acid is added to the fat phase and/or potassium sorbate to the water phase in appropriate quantities.

Another significant commercial application is the use of sorbates for the preservation of mayonnaise and delicatessen products containing mayonnaise as an ingredient. As oil-in-water emulsions, these mayonnaise products are more susceptible to microbial attack than emulsions of the opposite type. Commercially, mixtures of potassium sorbate and sodium benzoate have been introduced because of the risk that lactic acid bacteria might cause spoilage in weakly acid products.

19.10.2 Dairy Produce

Cheeses of all kinds are one of the main fields of use for sorbic acid, its choice being dictated in this instance by its favorable action in the high pH range and its specific action on molds (Lück 1968). Sorbic acid and sorbates are used both to preserve hard cheese during the ripening period and to protect cheese in consumer packs. The action of sorbic acid against mycotoxin-forming microorganisms is especially important in this connection (Wallhäußer and Lück 1970).

Sorbic acid is applied essentially by the following methods, the choice depending on the type of cheese and aim of the preservation:

1) in the form of sorbic acid or potassium sorbate as an addition to fresh or processed cheese;
2) in the form of potassium sorbate as an addition to the brine;
3) in the form of sorbic acid powder for the dusting of cheese;
4) in the form of aqueous sorbate solution for the dipping, spraying or washing of cheese;
5) in the form of calcium sorbate suspension for the treatment of ripening hard cheese;
6) in the form of fungistatic packaging materials or coating compounds containing sorbic acid, potassium sorbate or calcium sorbate as active ingredients.

When added to cheese, e.g. fresh or processed cheese, sorbic acid is generally used in concentrations of 0.05 to 0.07% (Perry and Lawrence 1960, Glandorf and Lück 1969). For surface treatment of cheese during ripening, sorbic acid concentrations of 0.1 to 0.4 g/dm^2 are required. The concentration of sorbic acid for fungistatic packaging materials is 2–4 g per m^2 (Smith and Rollin 1954, Lück 1962).

19.10.3 Meat Products

Undesired mold growth on hard sausages and frankfurter type sausages can be suppressed by treatment with 10–20% potassium sorbate solution (Leistner et al. 1975, Holley 1981). Attempts have also been made to protect beef (Lück 1984, Kondaiah et al. 1985) and poultry (Lück 1984, Robach et al. 1980, To and Robach 1980, Robach and Sofos 1982) against the growth of putrefactants and toxin-forming microorganisms by dipping them in 5 to 10% potassium sorbate solution. When used in conjunction with appropriate cooling and vacuum packing this has considerably increased storage stability.

There has been some discussion on the possibility of employing sorbic acid instead of, or in combination with, reduced quantities of nitrite to control clostridia and other toxin-forming bacteria in cured meat (Sofos et al. 1979, Robach and Sofos 1982, Lück 1984). The opinion, based on nutrient medium tests, used to be that sorbic acid was largely ineffectual against these microorganisms, but it has since been discovered that combinations of sorbates with small quantities of nitrite and/or phosphate in meat products at pH values of about 6 will inhibit clostridia as well as their toxin formation and other bacteria just as effectively as, or even better than, the nitrite concentrations hitherto used (Ivey and Robach 1978, Leistner et al. 1978, Ivey et al. 1978, Sofos et al. 1979, Robach and Sofos 1982). In addition, sorbic acid inhibits the formation of some nitrosamines in vitro (Tanaka et al. 1978). However, the use of sorbic acid and potassium sorbate as a "substitute" for nitrite has not become established commercially because sorbic acid, unlike nitrite, does not cause reddening of the meat nor does it help to produce a cured flavor.

19.10.4 Fish Products

Used in conjunction with other measures, such as salting, cooling and vacuum packing, sorbic acid has a good antibacterial action on fresh fish. It therefore also reduces the formation of trimethylamine and other undesired odors as well as the growth of pathogenic microorganisms (Thakur and Patel 1994). As it is highly effective against molds, sorbic acid is employed to preserve dried fish, such as klip-fish, which are susceptible to mold attack. The use of sorbic acid in lightly salted oriental fish preparations is of major commercial importance.

19.10.5 Vegetable Products

Sorbic acid is used in the form of water-soluble sorbates for preserving both fermented vegetable products and vegetables pickled in vinegar. When the sorbates are used to counter spoilage of vegetable products, the relatively weak action of sorbic acid against lactic acid bacteria is an advantage. If 0.05 to 0.15% potassium sorbate is added to the vegetable products prior to fermentation, the quantity of potassium sorbate depending on the product's salt content, the desired lactic acid fermentation is inhibited only slightly, if at all, whereas sorbic acid suppresses the growth of undesired film-forming yeasts and molds, thereby ensuring a "clean" lactic acid fermentation (Lück 1966). The yield for pickles can thus be improved by 20% (Courtial 1968). Gherkins and olives in vinegar are protected by 0.1 to 0.2% potassium sorbate against film-forming yeasts and mold attack. Another application of great commercial importance is the use of potassium sorbate in the preservation of oriental fermented vegetable products and spice sauces (Kato and Yoshii 1966). In the preservation of tomato products, sorbic acid is frequently combined with common salt and/or vinegar.

19.10.6 Fruit Products

Sorbic acid is used in a concentration of 0.05% for preserving ready-to-eat dried prunes produced by rehydration of more extensively predried fruit. Owing to their water activity, these prunes are susceptible only to mold attack (Nury et al. 1960). Fruit pulps can be protected against fermentation and mold attack by the addition of 0.1 to 0.13% potassium sorbate. However, as sorbic acid has no effect against oxidation or enzymatic spoilage, it is combined with small quantities of sulfur dioxide when used in such products. In jams, marmalades and jellies, the high sugar content means that a quantity of 0.05% sorbic acid is usually adequate for preservation purposes. Frequently, its use is confined to a surface treatment of the packaged products. In some countries sorbic acid is also important as a preservative for home-made jams.

19.10.7
Drinks

The remarks in the foregoing concerning fruit pulps also apply to the preservation of pure fruit juices. Sorbic acid is also used in the form of potassium sorbate mainly to preserve fruit juices destined for further processing. In general, potassium sorbate is combined with small quantities of SO_2 in order to protect the products additionally against oxidation, enzymatic and bacterial spoilage (lactic acid and acetic acid fermentation). In addition, pasteurization is carried out to inactivate the enzymes and reduce the microbe count. The applied concentration of potassium sorbate is 0.05 to 0.2%, depending on the nature of the juices and the desired keeping-time. In the case of soft drinks potassium sorbate is used in a dosage of 0.02% as an additional protection against spoilage by yeast.

Sorbic acid is highly important in all wine-growing countries, where it is used to stabilize wine against refermentation, since the normal concentrations of sulfur dioxide in wine have only a low efficacy against yeasts and afford no protection against the refermentation they cause. A combination of 200 mg sorbic acid, corresponding to 270 mg potassium sorbate per liter and some 20 to 40 mg free SO_2, provides the wine with good and comprehensive protection. Wines destined for stabilization with sorbic acid should be rendered as free from microorganisms as possible by suitable measures. As sorbic acid does not protect the wine against enzymatic changes, bacterial fermentation and oxidation, the conjoint use of sulfur dioxide is essential. Since the spectrum of action of sorbic acid differs from that of sulfur dioxide, the former cannot be regarded as a substitute for the latter. A certain quantity of sulfur dioxide is necessary, amongst other things, in order to suppress the lactic acid bacteria, some strains of which are able to reduce sorbic acid to sorbinol, which can chemically react with ethanol to form 1-ethoxy-2,4-hexadiene and 2-ethoxy-3,5-hexadiene (Crowell and Guymon 1975, von Rymon Lipinski et al. 1975). The latter compound, which has a powerful "flowery" scent, is the cause of the "geranium off-odor" in wine improperly treated with sorbic acid. If lactic acid bacteria are inhibited by the correct use of sulfur dioxide, this defect in the wine is prevented.

19.10.8
Baked Goods

Like propionic acid, which is widely used for the preservation of baked goods, sorbic acid is still effective at a high pH. Compared with propionates, sorbic acid is notable amongst other things for having a considerably more powerful antimicrobial action (Melnick et al. 1956), especially against *Trichosporon variabile*, a mold occasionally found on rye bread (Lamprecht 1955). Sorbic acid is used in quantities of 0.1 to 0.2%, relative to the flour, and is added to the dough while this is being prepared.

Sorbic acid is employed in baked goods, especially bread, not only for economic reasons but also on account of its action against aflatoxin-forming molds (Wallhäußer and Lück 1970). The use of sorbic acid in baked goods creates no problems, provided that baking powder is used to raise the dough, e.g. in cakes and other fine baked goods or pastries. Sorbic acid is added to the dough in this instance in concentrations of 0.1 to 0.2%, depending on the nature of the product and the desired keeping-time. The efficacy of sorbic acid against yeasts may adversely affect the leavening of bread. If so, its inhibitory action has to be compensated for by increasing the quantities of yeast and/or extending the leavening times. The mixed anhydride of sorbic acid and palmitic acid, sorboyl palmitate, has no antimicrobial action of its own but is split chemically in the baking process to release sorbic acid, which protects the finished bread against mold (Neu 1973). Sorboyl palmitate does not inhibit leavening at all, but has not become established commercially on account of its price. As an alternative, a form of sorbic acid has been developed which dissolves very slowly during preparation of the dough and does not affect the leavening process but still displays its full effect in the finished bread (Lück and Remmert). These characteristics are achieved by using a specific particle size. The product is marketed under the name Panosorb.

19.10.9 Confectionery

As sorbic acid is very effective in the high pH range of many types of confectionery, besides having a powerful action against osmophilic yeasts and also a neutral flavor, it is of some commercial importance in the preservation of fillings for chocolate and pralines. The applied concentration for this purpose is between 0.05 and 0.2%, according to the products' content of sugar and acids or other factors influencing the preserving action.

19.11 General Literature

Keller CL, Balaban SM, Hickey CS and DiFate VG (1983) Sorbic acid. In Kirk-Othmer: Encyclopedia of chemical technology, Volume 21, 3rd edition. John Wiley, New York, p. 402–416
Lück E (1969 –1973) Sorbinsäure. Chemie – Biochemie – Mikrobiologie –Technologie – Recht. (4 volumes) B Behr's Verlag, Hamburg
Lück E (1990) Food applications of sorbic acid and its salts. Food Add Contam 7, 711–715
Lück E (1993) Sorbic acid. In: Ullmann's Encyclopedia of Industrial Chemistry, Volume A 24, 5th edition. VCH Publishers, Weinheim, p. 507–513
Sofos JN(1989) Sorbate food preservatives. CRC Press, Boca Raton

19.12 Specialized Literature

Azukas JJ (1962) Sorbic acid inhibition of enolase from yeast and lactic acid bacteria. Thesis Michigan State Univ
Becker E, Roeder I (1957) Sorbinsäure als Konservierungsmittel für Margarine. Fette, Seifen, Anstrichm 59, 321–328
BIBRA, British Industrial Biological Research Organization (1987) Toxicity Profile, Sorbic acid and its common salts Carshalton
Bui L, Kooper C (1987) Reversed-phase liquid chromatographic determination of benzoic and sorbic acids in foods. J Assoc Off Anal Chem 70, 892–896

Courtial W (1968) Kaliumsorbat und sein Einsatz bei der Herstellung von Salzgurken. Ind Obst- u Gemüseverwertung 53, 381–383
Cox P (1994) Sorbic acid. Acta Cryst C 50, 1620–1622
Cremer HD, Tolckmitt W, Wenderhold J (1959) Beitrag zur Physiologie der Sorbinsäure. Klin Wochenschr 37, 304
Crowell EA, Guymon JF (1975) Wine constituents arising from sorbic acid addition, and identification of 2-ethoxyhexa-3,5-diene as source of geranium-like off-odor. Am J Enol Vitic 26, 97–102
Demaree GE, Sjogren DW, McCashland BW, Cosgrove FP (1955) Preliminary studies on the effect of feeding sorbic acid upon the growth, reproduction and cellular metabolism of albino rats. J Am Pharm Assoc Sci Ed 44, 619–621
Deuel HJ, Alfin-Slater R, Weil CS, Smyth HF (1954 a) Sorbic acid as a fungistatic agent for foods. I. Harmlessness of sorbic acid as a dietary component. Food Res 19, 1–12
Deuel HJ, Calbert CE, Anisfeld L, McKeehan H, Blunden HD (1954 b) Sorbic acid as a fungistatic agent for foods. II. Metabolism of a,b-unsaturated fatty acids with emphasis on sorbic acid. Food Res 19, 13–19
Dickens F, Jones HEH, Waynforth HB (1968) Further tests on the carcinogenicity of sorbic acid in the rat. Brit J Cancer 22, 762–768
Eklund T (1980) Inhibition of growth and uptake processes in bacteria by some chemical food preservatives. J Appl Bacteriol 48, 423–432
Eklund T (1981) Chemical food preservation – some basic aspects and practical considerations. Applied food science in food preservation (paper presented in London on November 18, 1981)
Eklund T (1983) The antimicrobial effect of dissociated and undissociated sorbic acid at different pH levels. J Appl Bacteriol 54, 383–389
Eklund T (1984) The antimicrobial action of some food preservatives at different pH levels. In: Kiss I, Deák T and Incze K: Microbial Associations and Interactions in Food. Budapest p. 441–445
Eklund T (1985) The effect of sorbic acid and esters of
p-hydroxybenzoic acid on the protonmotive force in *Escherichia coli* membrane vesicles. J Gen Microbiol 131, 73–76
Emard LO, Vaughn RH (1952) Selectivity of sorbic acid media for the catalase negative lactic acid bacteria and clostridia.
J Bacteriol 63, 487–494
Fingerhut M, Schmidt B, Lang K (1962) Über den Stoffwechsel der 1-^{14}C-Sorbinsäure. Biochem Z 336, 118–125
Finol ML, Marth EH, Lindsay RC (1982) Depletion of sorbate from different media during growth of *Penicillium* species. J Food Protect 45, 398–404
Flak W, Schaber R (1988) Die Bestimmung von Konservierungsmitteln in Wein und anderen Getränken mittels
Hochdruckflüssigkeitschromatographie. Mitt Klosterneuburg 38, 10–16
Food and Drug Research Laboratories, Inc (1973) Scientific literature reviews on Generally Recognized as Safe (GRAS) food ingredients – sorbic acid and derivatives. PB-223 864. National Technical Information Service. US Department of Commerce, Springfield
Food and Drug Research Laboratories, Inc (1975) Teratologic evaluation of FDA 73–4, potassium sorbate: Sorbistat in mice and rats. PB-245 520. National Technical Information Service. US Department of Commerce, Springfield
Gaunt IF, Butterworth KR, Hardy J, Gangolli SD (1975) Long-term toxicity of sorbic acid in the rat. Food Cosmet Toxicol 13, 31–45
Glandorf K, Lück E (1969) Sorbinsäure als Konservierungsstoff für Schmelzkäse. Molk Ztg (Hildesheim) 23, 1043–1044
Häberle M (1989) Pseudoallergische Reaktion auf Konservierungs- und Farbstoffe. Ernährungs-Umschau 36, 8–16
Hagenauer-Hener U, Frank C, Hener U, Mosandl A (1990) Bestimmung von Aspartam, Acesulfam-K, Saccharin, Coffein, Sorbinsäure und Benzoesäure in Lebensmitteln mittels HPLC. Dtsch Lebensm Rundsch 83, 348–351

Hannuksela M, Haahtela T (1987) Hypersensitivity reactions to food additives. Allergy 42, 561–575

Hendy RJ, Hardy J, Gaunt IF, Kiss IS, Butterworth KR (1976) Long-term toxicity studies of sorbic acid in mice. Food Cosmet Toxicol 14, 381–386

Holley RA (1981) Prevention of surface mold growth on Italian dry sausage by natamycin and potassium sorbate. Appl Environm Microbiol 41, 422–429

Ivey FJ, Robach MC (1978) Effect of sorbic acid and sodium nitrite on *Clostridium botulinum* outgrowth and toxin production in canned comminuted pork. J Food Sci 43, 1782–1785

Ivey FJ, Shaver KJ, Christiansen LN, Tompkin RB (1978) Effect of potassium sorbate on toxinogenesis by *Clostridium botulinum* in bacon. J Food Protect 41, 621–625

Jung R, Cojocel C, Müller W, Böttger D, Lück E (1992) Evaluation of the genotoxic potential of sorbic acid and potassium sorbate. Food Chem Toxicol 30, 1–7

Kantasubrata J, Imamkhasani S (1991) Analysis of additives in fruit juice using HPLC. ASEAN Food J 6, 155–158

Karovicova J, Polonski J, Simko P (1991) Determination of preservatives in some food products by capillary isotachophoresis. Nahrung 35, 543–544

Kato H, Yoshii H (1966) Studies on the microbiology of vegetables soaked in soy-sauce. Part II. Sorbic acid tolerance of principal strains. Nippon Shokuhin Kogyo Gakkai-Shi 13, 137–140

Kinderlerer J, Hatton P (1990) Fungal metabolites of sorbic acid. Food Add Contam 7, 657–669

Koch CS (1994) Ökotoxikologische Daten von Kosmetik-Konservierungsstoffen. Seifen-Öle-Fette-Wachse. J 120, 655–660

Kondaiah N, Zeuthen P, Jul M (1985) Effect of chemical dips on unchilled fresh beef inoculated with *E. coli, S. aureus, S. faecalis* and *Cl. perfringens* and stored at 30 °C and 20 °C. Meat Sci 12, 17–30

Küppers S (1988) Reversed-phase liquid chromatographic determination of benzoic and sorbic acids in fresh cheese. J Assoc Off Anal Chem 71, 1068–1071

Kuhn R, Köhler F, Köhler L (1937) Über die biologische Oxydation hochungesättigter Fettsäuren. Ein neuer Weg zur Darstellung von Polyendicarbonsäuren. Hoppe Seyler's Z Physiol Chem 247, 197–219

Lamprecht F (1955) Zur Frage der Bekämpfung des Brotschimmels insbesondere des Kreideschimmels. Brot Gebäck 9, 26–30

Lang K (1960) Die Verträglichkeit der Sorbinsäure. Arzneim Forsch 10, 997–999

Leistner L, Maing IY, Bergmann E (1975) Verhinderung von unerwünschtem Schimmelpilzwachstum auf Rohwurst durch Kaliumsorbat. Fleischwirtschaft 55, 559–561

Leistner L, Bem Z, Hechelmann H (1978) Potassium sorbate - an alternative to nitrite in meat products. Proc 24th Europ Meat Res Congr 1978 III W 2: 1–9

Liewen MB, Marth EH (1985) Growth and inhibition of microorganisms in the presence of sorbic acid: A review. J Food Protect 48, 364–375

Litton Bionetics, Inc (1974) Mutagenic evaluation of compound FDA 73–4, potassium sorbate. PB-245 434. National Technical Information Service. US Department of Commerce. Springfield

Litton Bionetics, Inc (1977) Mutagenic evaluation of compound FDA 75-73. 00749Z-55-9. Calcium sorbate. PB-266 894. Technical Information Service. US Department of Commerce, Springfield

Luckmann FH, Melnick D (1955) Sorbic acid as a fungistatic agent for foods. X. Spectrophotometric determination of sorbic acid in foods in general. Food Res 20, 649–654

Lück E (1962) Fungistatische Verpackungsmaterialien auf Basis Sorbinsäure und Calciumsorbat. Dtsch Lebensm Rundsch 50, 353–357

Lück E (1966) Sorbinsäure und Kaliumsorbat als Hilfsmittel bei der Herstellung milchsauer vergorener Gurken. Ind Obst- u Gemüseverwertung 51, 410–414

Lück E (1968) Sorbinsäure als Konservierungsstoff für Käse. Dtsch Molk Ztg 89, 520–521

Lück E (1969) Sorbinsäure. Chemie – Biochemie – Mikrobiologie – Technologie. Volume 1. B. Behr's Verlag, Hamburg, p. 109–123

Lück E (1972) Sorbinsäure. Chemie – Biochemie – Mikrobiologie –Technologie. Volume 2. B. Behr's Verlag, Hamburg, p. 15–18

Lück E (1972) Sorbinsäure. Chemie - Biochemie - Mikrobiologie -Technologie. Volume 2. B. Behr's Verlag, Hamburg, p. 21

Lück E (1972) Sorbinsäure. Chemie - Biochemie - Mikrobiologie -Technologie. Volume 2. B. Behr's Verlag, Hamburg, p. 42-87

Lück E (1984) Sorbinsäure und Sorbate. Konservierungsstoffe für Fleisch und Fleischwaren. Literaturübersicht. Fleischwirtschaft 64, 727-733

Lück E (1993) Sorbic Acid. In: Ullmanns Encyclopedia of Industrial Chemistry, Volume A24, 5th edition. VCH Publishers, Weinheim, p. 507-513

Lück E, Remmert K-H (Hoechst Aktiengesellschaft) Verfahren zum Konservieren von mit Hefe und/oder Sauerteig getriebenen Backwaren mit Sorbinsäure. Europ Patent 75-286

Lück H (1957) Katalasehemmung durch organische Säuren. Biochem Z 328, 411-419

Lück H (1958) Einflub von Konservierungsmitteln auf Häminenzyme. I. Mitteilung. Allgemeine Wirkung. Z Lebensm Unters Forsch 108, 1-9

Lück H, Rickerl E (1959) Untersuchungen an *Escherichia coli* über eine Resistenzsteigerung gegen Konservierungsmittel und Antibiotica. Z Lebensm Unters Forsch 109, 322-329

Lukas E-M (1964) Zur Kenntnis der antimikrobiellen Wirkung der Sorbinsäure II. Mitteilung. Die Wirkung der Sorbinsäure auf *Aspergillus niger* von Tieghem und andere Schimmelpilze. Zentralbl Bakteriol Parasitenkd Infektionskr Hyg, II. Abt 117, 485-509

Marth EH, Capp CM, Hasenzahl L, Jackson HW, Hussong RV (1966) Degradation of potassium sorbate by *Penicillium* species. J Dairy Sci 49, 1197-1205

Martoadiprawito W, Whitaker JR (1963) Potassium sorbate inhibition of yeast alcohol dehydrogenase. Biochem Biophys Acta 77, 536-544

Melnick D, Vahlteich HW, Hackett A (1956) Sorbic acid as a fungistatic agent for foods. XI. Effectiveness of sorbic acid in protecting cakes. Food Res 21, 133-146

Münzner R, Guigas C, Renner H (1990) Re-examination of potassium sorbate and sodium sorbate for possible genotoxic potential. Food Chem Toxicol 28, 397-401

Neu H (1973) Die Schimmelverhütung bei Brot durch Sorbinsäurepalmitinsäureanhydrid (Sorboylpalmitat). Dtsch Lebensm Rundsch 69, 401-404

Nury FS, Miller MW, Brekke JE (1960) Preservative effect of some antimicrobial agents on high-moisture dried fruits. Food Technol 14, 113-115

Oka S (1960) Studies on transfer of antiseptics to microbes and their toxic effect. Part I. Accumulation of acid antiseptics in yeast cells. Bull Agric Chem Soc Japan 24, 59-65

Olea Serrano F, Lopes I, Revilla N (1991) High performance liquid chromatography determination of chemical preservatives in yogurt. J Liq Chromat 14, 709-717

Perry GA, Lawrence RL (1960) Preservative effect of sorbic acid on creamed cottage cheese. J Agric Food Chem 8, 374-376

Radler F (1976) Dégradation de l'acide sorbique par les bactéries. Bull OIV 49, 629-635

Rehm H-J (1961) Grenzhemmkonzentrationen der zugelassenen Konservierungsmittel gegen Mikroorganismen. Z Lebensm Unters Forsch 115, 293-309

Rehm H-J (1967) Zur Kenntnis der antimikrobiellen Wirkung der Sorbinsäure. 6. Mitteilung. Die Wirkung von Sorbinsäure auf den Kohlenhydratstoffwechsel von *Escherichia coli*. Zentralbl Bakteriol Parasitenkd Infektionskr Hyg Abt 2, 121, 492-502

Rehm H-J, Lukas E-M (1963) Zur Kenntnis der antimikrobiellen Wirkung der Sorbinsäure. 1. Mitteilung. Die Wirkung der undissoziierten und dissoziierten Anteile der Sorbinsäure auf Mikroorganismen. Zentralbl Bakteriol Parasitenkd Infektionskr Hyg, II. Abt 117, 306-318

Rehm H-J, Wallnöfer P (1964) Zur Kenntnis der antimikrobiellen Wirkung der Sorbinsäure. Naturwissenschaften 51, 13-14

Reifschneider C, Klug C, Jager M (1994) Identifizierung und Quantifizierung von Konservierungsstoffen in kosmetischen Mitteln. Seifen-Öle-Fette-Wachse J 120, 650-654

Robach MC, To EC, Meydav S, Cook CF (1980) Effect of sorbates on microbiological growth in cooked turkey products. J Food Sci 45, 638-640

Robach MC, Sofos JN (1982) Use of sorbates in meat products, fresh poultry and poultry products: A review. J Food Protect 45, 374-383

Rosenhall L (1982) Evaluation of intolerance to analgesics, preservatives and food colorants with challenge tests. Eur J Respir Dis 63, 410-419

Rymon Lipinski G-W von, Lück E, Oeser H, Lömker F (1975) Entstehung und Ursachen des "Geranientons". Mitt Rebe, Wein, Obstbau Früchteverwert 25, 387–394
Sado I (1973) Synergistic toxicity of official permissible preservative food additives. Nippon Eiseigaku Zasshi 28, 463–476
Schiffmann D, Schlatter J (1992) Genotoxicity and cell transformation studies with sorbates in Syrian hamster embryo fibroblasts. Food Chem Toxicol 30, 669–672
Schlatter J, Würgler F, Kränzlin R, Maier P, Holliger E, Graf U (1992) The potential genotoxicity of sorbates: effects on cycle in vitro in V 79 cells and somatic mutations in drosophila. Food Chem Toxicol 30, 843–851
Schmidt H (1960) Eine spezifische colorimetrische Methode zur Bestimmung der Sorbinsäure. Z Anal Chem 178, 173–184
Schweizer Lebensmittelbuch (Swiss Foodstuffs Manual), Chapter 44: "Konservierungsmittel für Lebensmittel", New edition, 1st instalment, May 1992, Bundesamt für das Gesundheitswesen, Bern
Shtenberg AJ, Ignat'ev AD (1970) Toxicological evaluation of some combinations of food preservatives. Food Cosmet Toxicol 8, 369–380
Smith DP, Rollin NJ (1954) Sorbic acid as a fungistatic agent for foods. VII. Effectiveness of sorbic acid in protecting cheese. Food Res 19, 59–65
Smyth HF, Carpenter CP (1948) Further experience with the range finding test in the industrial toxicological laboratory. J Ind Hyg Toxicol 30, 63–68
Sofos J (1992) Sorbic acid, mode of action. In: Encyclopedia of microbiology. J Lederberg (ed) Academic Press New York
Sofos JN, Busta FF (1981) Antimicrobial activity of sorbate. J Food Protect 44, 614–622
Sofos JN, Busta FF, Allen CE (1979) Botulism control by nitrite and sorbate in cured meats: A review. J Food Protect 42, 739–770 and 1099
Tanaka K, Chung KC, Hayatsu H, Kada T (1978) Inhibition of nitrosamine formation in vitro by sorbic acid. Food Cosmet Toxicol 16, 209–215
Thakur BR, Patel TR (1994) Sorbates in fish and fish products – A review. Food Rev Int 10, 93–107
Thakur BR, Singh RK, Arya SS (1994) Chemistry of sorbates – A basic perspective. Food Rev Int 10, 71–91
To EC, Robach MC (1980) Inhibition of potential food poisoning microorganisms by sorbic acid in cooked, uncured, vacuum packaged turkey products. J Food Technol 15, 543–547
Vieths S, Fischer K, Dehne LI, Bögl KW (1994) Allergenes Potential von verarbeiteten Lebensmitteln. Ernährungs-Umschau 41, 140–143 and 186–190
Viñas I, Morlans I, Sanchis V (1990) Potential for the development of tolerance by *Aspergillus amstelodami, A. repens* and *A. ruber* after repeated exposure to potassium sorbate. Zbl Mikrobiol 145, 187–193
Walker R (1990) Toxicology of sorbic acid and sorbates. Food Add Contam 7, 671–676
Wallhäußer KH, Lück E (1970) Der Einfluß der Sorbinsäure auf mycotoxinbildende Pilze in Lebensmitteln. Dtsch Lebensm Rundsch 66, 88–92
Westöö G (1964) On the metabolism of sorbic acid on the mouse. Acta Chem Scand 18, 1373–1378
Würgler F, Schlatter J, Maier P (1992) The genotoxicity status of sorbic acid, potassium sorbate and sodium sorbate. Mutat Res 283, 107–111
York GK (1950) Studies on the inhibition of microbes by sorbic acid. Thesis Univ California
York GK, Vaughn RH (1955) Resistance of *Clostridium parabotulinum* to sorbic acid. Food Res 20, 60–65
York GK, Vaughn RH (1964) Mechanisms in the inhibition of microorganisms by sorbic acid. J Bacteriol 88, 411–417

Chapter 20

Dicarbonic Acid Esters

20.1 Synonyms

IUPAC: Dicarbonic Acid Esters
English: Pyrocarbonic acid esters. *German:* Dikohlensäureester, Pyrokohlensäureester. *French:* Esters d'acide dicarbonique. *Italian:* Esteri dell'àcido dicarbonico. *Spanish:* Esteres del acido dicarbónico. *Russian:* Эфиры пироуголъной кислоты.

20.2 History

The antimicrobial action of pyrocarbonic acid esters was first described in 1956 by H. Bernhard, W. Thoma and H. Genth. By the mid-1960s, dicarbonic acid diethyl ester had become important in a number of countries for the preservation of wine and soft drinks, but in about 1973 it was withdrawn for toxicological reasons. The dimethyl ester, dimethyl dicarbonate, is now used instead.

20.3 Commercially Available Forms, Derivatives

The products of interest from the preservation viewpoint are low alkyl esters of dicarbonic acid. The diethyl ester (DEPC), known in the Federal Republic of Germany by the abbreviation PKE and the trade name Baycovin, was the first to be used. Recently the dimethyl ester, also known as DMDC and by the trade name Velcorin, has been introduced.

20.4 Properties

CH_3O–CO–O–CO–OCH_3, molar mass 134.09, colorless, transparent liquid with a fruity odor, solidifies at 17 °C. Dicarbonic acid dimethyl ester is miscible with ethanol and other alcohols, and will dissolve in water up to 3.65 %.

C_2H_5O–CO–O–CO–OC_2H_5, molar mass 162.14, colorless, transparent liquid with a fruity odor, boils at 180 °C upon decomposing. Dicarbonic acid diethyl ester is miscible with ethanol and other alcohols, and will dissolve in water up to 0.6 %.

In the presence of water both esters hydrolyze to methanol and ethanol respectively and carbon dioxide. The hydrolysis is independent of the CO_2 pressure but nevertheless governed by the temperature and pH value. The ethyl ester hydrolyzes completely in drinks at room temperature within 20 to 24 hours (Pauli and Genth 1966). The hydrolysis of the methyl ester proceeds under the same conditions but three to four times as rapidly (Genth 1979). Dicarbonic acid dimethyl ester is broken down to its detection limit in highly acidic fruit juices with pH values around 2.8 at temperatures ranging between 10 and 30 °C within 65 to 260 min.

Apart from reacting with water the esters also react with other constituents of drinks. With ethanol, they form diethyl carbonate or methylethyl carbonate (Ough and Langbehn 1976, Stafford and Ough 1976). In the presence of ammonium salts, they form ethyl carbamate from the ethyl ester (Löfroth and Gejvall 1971, Ough 1976, Solymosy et al. 1978) or methyl carbamate from the methyl ester (Genth 1979), each in small quantities. Dimethyl dicarbonate also reacts with higher alcohols, though the quantities likely to be found in wine are in the ppb range (Peterson and Ough 1979).

20.5 Analysis

To detect whether wine has been treated with dicarbonic acid diethyl ester, a gas chromatographic test can be conducted on the wine to check for the presence of diethyl carbonate (Kiehlhöfer and Würdig 1963). Methyl ethyl carbonate can also be detected by gas chromatography (Pauli and Genth 1966).

The dicarbonates also can be detected by photometry and titrimetry (Ough 1993). A gas chromatographic procedure based on extraction with carbon disulfide is the official analysis method of the AOAC (Horwitz 1980).

20.6 Health Aspects

20.6.1 Acute Toxicity

The LD_{50} of dicarbonic acid diethyl ester administered to rats in an oily solution is 1–1.5 g/kg body weight (Hecht 1961). For the methyl ester the acute toxicity for rats after oral administration, at 350–500 mg/kg body weight, is very high because of local irritant effects (Classen et al. 1987). For substances that hydrolyze rapidly in drinks, the acute toxicity is of less importance than the toxicological assessment of reaction products that may occur in drinks containing these substances.

Both dicarbonic acid diethyl ester and dicarbonic acid dimethyl ester cause irritation of the skin and mucous membranes. Consequently they must be handled with care.

20.6.2 Subchronic Toxicity

When freshly prepared oily solutions of dicarbonic acid diethyl ester were administered to rats over a period of several months, no untoward effects were apparent (Hecht 1961). Furthermore, drinks containing an addition of 4 g dimethyl dicarbonate per liter were tolerated without reaction over 3 months (Genth 1979).

20.6.3 Chronic Toxicity

The most relevant factor in any assessment of the chronic toxicity of dicarbonic acid diethyl ester is its reaction product, ethyl carbamate (Löfroth and Gejvall 1971). It has a carcinogenic effect, after both oral and inhalative exposure, on rats, mice and hamsters.

Ethyl carbamate is classified by the Federal German MAK committee in list III A 2 as being definitely carcinogenic in animal experiments (Schmähl et al. 1979, Senatskommission zur Prüfung gesundheitsschädlicher Arbeitsstoffe 1993). Its carcinogenic effect is probably based on its oxidation to epoxyethyl carbamate via vinyl carbamate as an intermediate stage. Corresponding oxoethyl adducts to DNA and proteins have been detected (Park et al. 1993). It was the carcinogenic effect of the ethyl carbamate formed from the reaction of dicarbonic acid diethyl ester with ammonium ions that led to the prohibition of this substance.

The corresponding reaction product of dicarbonic acid dimethyl ester, methyl carbamate, is not carcinogenic, however (Genth 1979) and is regarded as safe (Classen et al. 1987). Methyl carbamate is permitted in drinks in concentrations of up to 0.02 mg/l. The SCF has no objections on toxicological grounds to the use of dicarbonic acid dimethyl ester in concentrations of 125 to 250 mg/l in soft drinks (SCF 1990). The toxicological data are summarized in a report by JECFA (JECFA 1991).

20.7 Regulatory Status

The approvals granted in some countries for dicarbonic acid diethyl ester have now all been withdrawn. In a number of countries there have been times when no approval has been needed for the methyl ester, dimethyl dicarbonate as an additive in soft drinks, for example, though not in wine, because it is not present in the food or drink at the time of consumption. In the European Union it will in future be permitted, as E 242, in certain drinks. Limited approvals exist in other countries as well, e.g. in the US for wine, iced tea and tea extracts.

20.8 Antimicrobial Action

20.8.1 General Criteria of Action

Dicarbonic acid esters kill microorganisms faster than do conventional preservatives. Consequently, the nature of their action classifies them as disinfectants rather than preservatives. Because of their special properties the products are also termed chemical sterilants, sterilization auxiliaries or cold sterilization agents.

The antimicrobial action of dicarbonic acid esters is based on their reaction with enzymes inside the microorganisms and with membranes (Genth 1964).

The dosage of dicarbonic acid esters necessary to kill microorganisms depends to a large extent on the microbe count. An increase in the microbe count by a power of ten necessitates doubling the dosage needed to confer good keeping properties (Table 27).

Table 27. Action of dicarbonic acid esters against *Saccharomyces cerevisiae* in conjunction with various microbe counts

Microbe count per ml	Concentration in ppm required to kill the microorganisms
6000	200
600	100
60	60
< 10	30

20.8.2 Spectrum of Action

Dicarbonic acid esters act primarily against yeasts. Their effect against bacteria is far weaker, and that against molds the weakest of all. Owing to the marked dependence of their action on the microbe count, the data given in the literature on the minimum inhibitory concentration of the substances against the individual types of microorganism are not always comparable.

Most yeasts at low microorganism counts (400–600 per ml) are killed by concentrations of some 100 ppm. In model experiments with wine, most yeast strains were killed in levels of 50 to 200 cells per ml by concentrations of as little as 25 mg DMDC/l wine (Dauth and Ough 1980). Bacteria and molds require dosages of 3 to 5 times this level (Genth 1979).

20.9 Fields of Use

The possible uses of the dicarbonic acid esters are determined by their unique property among preservatives, namely that of decomposing to microbiologically inactive substances after a certain period of time in the presence of water. Hence, the low alkyl esters of dicarbonic acid can be used only for products in which the microorganisms present are killed before hydrolysis and which can then be protected from reinfection by technical measures. Because of these preconditions, the use of these products is confined to drinks which are transferred to sealed containers directly after treatment. In view of their rapid decomposition, especially in the case of the methyl ester, the addition must be made immediately before bottling.

To incorporate the preservative in the drinks, use is made of specially designed pumps which allow the small quantities added to be injected reliably and continuously into the drink. When dimethyl dicarbonate is used in cool rooms, the metering pumps must be designed to raise its temperature to 17 °C, since it solidifies below this point.

As both the esters have approximately the same efficacy, the recommended applied dosages are more or less the same. For carbonated soft drinks these amount to 50 – 150 ppm. For still drinks, 100 – 200 ppm are required. The dosage of dicarbonic acid diethyl ester formerly employed for wine was 110 – 170 ppm (Ough 1993).

Dicarbonic dimethyl ester can also be used as a yeast inhibitor in grape juices and tomato products (Anon 1993, Bizri and Wahem 1994). Owing to its rapid hydrolysis, however, it is advisable to use it in conjunction with potassium sorbate to prolong the protective effect, especially at high storage temperatures. At applied concentrations of around 250 ppm, dicarbonic dimethyl ester has an unfavorable effect in tomato juice on the color and on the content of vitamins such as β-carotin and vitamin C. This can be remedied by combining the dicarbonic dimethyl ester with potassium sorbate or sodium benzoate (Bizri and Wahem 1994).

20.10 Literature

Anon (1993) DMDC, a good grape juice inhibitor study shows. Food Chem News 22.11.1993, p. 10

Bernhard H, Thoma W, Genth J (Farbenfabriken Bayer, Leverkusen) (1956) Konservierungsmittel. Deutsches Patent 1 011 709 and US Patent 2 910 400

Bizri J, Wahem I (1994) Citric acid and antimicrobial effect, microbiological stability and quality of tomato juice. J Food Sci 59, 130–134

Classen H-G, Elias P, Hammes W (1987) Toxikologisch-hygienische Beurteilung von Lebensmittelinhalts- und -zusatzstoffen sowie bedenklicher Verunreingungen. Parey, Berlin, p. 120–122

Daudt CE, Ough CS (1980) Action of dimethyldicarbonate on various yeasts. Am J Enol Viticult 31, 21 – 23

Genth H (1964) On the action of diethylpyrocarbonate on microorganisms. Proc 4th Int Symp on Food Microbiol Göteborg, p. 77 – 85

Genth H (1979) Dimethyldicarbonat – ein neuer Verschwindestoff für alkoholfreie, fruchtsafthaltige Erfrischungsgetränke. Erfrischungsgetränk 13, 262 – 269

Hecht G (1961) Zur Toxikologie des Pyrokohlensäurediäthylesters. Z Lebensm Unters Forsch 114, 292 – 297

Horwitz W (1980) Official Methods of Analysis of the Association of Official Analytical Chemists, 13th edition. Washington: AOAC p. 192 – 193

JECFA (1991) WHO Food Additives Series No 28. Toxicological evaluation of certain food additives and contaminants. 37th meeting, p. 231 – 273

Kielhöfer E, Würdig G (1963) Nachweis und Bestimmung von Diäthylcarbonat und Pyrokohlensäurediäthylester im Wein und Schaumwein. Dtsch Lebensm Rundsch 59, 197 – 200

Löfroth G, Gejvall T (1971) Diethyl Pyrocarbonate: Formation of urethane in treated beverages. Science 174, 1248 – 1250

Ough CS (1976) Ethylcarbamate in fermented beverages and foods. II. Possible formation of ethylcarbamate from diethyl dicarbonate addition to wine. J Agric Food Chem 24, 328 – 331

Ough CS, Langbehn L (1976) Measurement of methylcarbamate formed by the addition of dimethyl dicarbonate to model solutions and to wines. J Agric Food Chem 24, 428 – 430

Ough CS (1993) Dimethyldicarbonate and diethyldicarbonate. In: Davidson M, Branen AL: Antimicrobials in Foods. Marcel Dekker, New York, p. 343–368
Park K-K, Liem A, Steward B, Müller J (1993) Vinyl carbamate epoxide, a major strong electrophilic, mutagenic and carcinogenic metabolite of vinyl carbamate and ethyl carbamate (urethane). Carcinogenesis 14, 441–450
Pauli O, Genth H (1966) Zur Kenntnis des Pyrokohlensäurediäthylesters. I. Mitteilung. Eigenschaften, Wirkungsweise und Analytik. Z Lebensm Unters Forsch 132, 216–227
Peterson TW, Ough CS (1979) Dimethyldicarbonate reaction with higher alcohols. Am J Enol Viticult 30, 119–123
SCF (1990) Bericht des wissenschaftlichen Lebensmittelausschusses der E.G., p. 10. EUR 139/13 DE
Schmähl D, Port R, Warendorf J (1977) A dose-response study on urethane carcinogenesis in rats and mice. Int J Cancer 19, 77–80
Senatskommission zur Prüfung gesundheitschädlicher Arbeitsstoffe (1993) Maximale Arbeitsplatzkonzentrationen und biologische Arbeitsstofftoleranzwerte. VCH Verlagsgesellschaft, Weinheim
Solymosy F, Antony F, Fedorcsák I (1978) On the amounts of urethane formed in diethyl pyrocarbonate treated beverages. J Agric Food Chem 26, 500–503
Stafford PA, Ough CS (1976) Formation of methanol and ethyl methyl carbonate by dimethyl dicarbonate in wine and model solutions. Am J Enol Viticult 27, 7–11

Chapter 21

Benzoic Acid

21.1
Synonyms

IUPAC: Benzoic acid
German: Benzoesäure. *French:* Acide benzoique. *Italian:* Acido benzoico. *Spanish:* Acido benzoico. *Russian:* Бензойная кислота.

21.2
History

The preservative action of benzoic acid was first described in 1875 by H. Fleck (Strahlmann 1974), when attempting to find a substitute for the already familiar salicylic acid. It was Fleck who established a relationship between the action of both these acids and that of phenol. Unlike salicylic acid, benzoic acid could not initially be produced synthetically in large quantities; so it was not until the turn of the century that it was first introduced for food preservation. Since then it has been a widely used preservative throughout the world, chiefly on account of its low price, although recently there has been a perceptible trend towards restricting its use in favor of other preservatives considered to be better from the toxicological viewpoint.

21.3
Commercially Available Forms

Benzoic acid is used as such and in the form of its sodium salt, sodium benzoate, which has better water solubility. In exceptional cases it is also employed as potassium benzoate.

21.4
Properties

C_6H_5COOH, molar mass 121.11, white, glossy monoclinic flakes or needles which melt at 122 °C. At room temperature 0.34 g benzoic acid dissolves in 100 g water and 1–2 g in 100 g fatty oils. Benzoic acid is readily soluble in anhydrous ethanol.

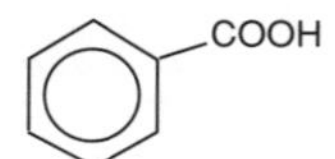

Sodium benzoate, molar mass 144.11, white crystalline powder with a water solubility at room temperature of 63 g/100 g.

21.5 Analysis

Owing to its volatility in steam, benzoic acid can be quantitatively isolated by acid steam distillation from the food being investigated. The sample can also be prepared by solid phase extraction (Moors et al. 1991).

After suitable preparation of the extracts, benzoic acid can be detected by spectrophotometry, HPLC or TLC methods and determined quantitatively (Khan et al. 1994). Most HPLC methods are based on separation of the preservatives on an RP phase followed by UV detection (Hagenauer-Hener et al. 1990, Reifschneider et al. 1994).

During the detection and quantitative determination of benzoic acid in foods, it should be borne in mind that it may occur naturally in free form or linked to glycosides in dairy produce and fruits, such as strawberries and cranberries (Sieber et al. 1990).

21.6 Production

Benzoic acid is produced on an industrial scale by catalytic oxidation or air oxidation of toluene. The hydrolysis of benzotrichloride and the treatment of molten phthalic acid anhydride with steam in the presence of zinc-containing catalysts are no longer of any industrial importance.

21.7 Health Aspects

21.7.1 Acute Toxicity

The LD_{50} of benzoic acid for rats after peroral administration is 1.7–3.7 g/kg body weight (Deuel et al. 1954, Sado 1973). Cats appear to be especially sensitive to benzoic acid, for as little as 0.3–0.6 g/kg body weight can prove fatal (Bedford and Clarke 1972). This is because cats lack the ability to convert benzoic acid enzymatically into hippuric acid.

21.7.2 Subchronic Toxicity

A daily dosage of 80 mg benzoic acid per kg body weight fed to mice over 3 months raises the mortality rate, especially when doses of sulfite are administrated at the same time (Shtenberg and Ignat'ev 1970). Within 4 to 5 days, 3% benzoic acid in the feed causes disturbances to the central nervous system, ataxia and tonic-clonic convulsions. However, there are no macroscopically visible changes to organs or histological evidence of disease in the heart, liver and kidneys. Half the experimental animals died after only 5 days and revealed necrotic changes in the brain. The administration of 1.1% benzoic acid resulted only in poorer weight development but no other damage (Kreis et al. 1967). In a further experiment 4% sodium benzoate in the diet fed for 90 days caused no damage (Deuel et al. 1954). In more recent feeding experiments with rats and mice involving 3% sodium benzoate relative to the feed over 10 days, there was an increase in kidney and liver weight, changes in the albumin and gamma-glutamyl-transpeptidase levels and hepatocytenic changes (Fujitani 1993).

Widely differing values are given in the literature for the toxicologically relevant threshold concentration of benzoic acid in subchronic tests (Informatics 1972, Federation of American Societies for Experimental Biology 1973).

In humans daily doses of 1 g benzoic acid for 90 days do not impair the health (Informatics 1972), nor does a daily intake of 12 g over 14 days (Waldo 1949) or 0.3 to 4 g over 60 to 100 days (Lang 1960).

Sodium benzoate has no teratogenic action after oral administration (Minor and Becker 1971, Food and Drug Research Laboratories 1972).

21.7.3 Chronic Toxicity

Doses of 40 mg benzoic acid per kg body weight per day administered to mice and rats over periods of 17 and 18 months, respectively, produced growth disturbances (Shtenberg and Ignat'ev 1970). In a feeding trial involving 40 rats, 5% sodium benzoate proved highly toxic and all the animals died within 2 weeks. A quantity of 1% was tolerated without inimical effects on growth, feed utilization, life span, reproduction, weight development or the histology of 6 organs (Lang 1960, Kieckebusch and Lang 1960). According to the findings of another author, the growth of rats whose diet contains 1.5% benzoic acid is markedly slower than that of control animals (Marquardt 1960).

Since 1979, benzoic acid has been used as a therapeutic agent for reducing ammonium toxicity in patients with enzyme defects in the urea cycle (Tremblay and Aureshi 1993). A side effect observed has been a reduction in the intramitochondrial *N*-Acetyl-glutamate level (O'Connor et al. 1989) of recent indications that benzoic acid may have clastogenic and teratogenic effects, the SCF has set a temporary ADI of 0–5 mg/kg body weight (SCF 1994). JECFA is retaining the previous ADI of 0–5 mg/kg body weight but demanding further toxicological investigations (JECFA 1993).

21.7.4 Intolerance Reactions

Like the esters of *p*-hydroxybenzoic acid and certain azo dyes, benzoic acid is a food additive with considerable sensitizing potential. After both oral administration and topical application benzoic acid can cause intolerant reactions such as urticaria, asthma and anaphylactic shock (Michils et al. 1991, Jacobsen 1992). A possible reason that has been discussed is a benzoate-induced release of histamine (Schaubschläger et al. 1991).

21.7.5
Biochemical Behavior

Benzoic acid is absorbed well from the intestine. It does not form links with protein. Instead benzoic acid is first converted with ATP into benzoyl-ATP and then, like fatty acids, "activated" by linkage with coenzyme A to yield benzoyl coenzyme A. Under the influence of glycin *N*-acylase, this, together with glycin, produces hippuric acid (benzoyl glycol), which is excreted in the urine (Lang 1960). In addition, relatively small quantities of benzoic acid are linked to glucuronic acid and excreted in the urine by this route (Trembley and Aureshi 1993).

21.8
Regulatory Status

In most countries benzoic acid and sodium benzoate have been permitted for food preservation for many years. Apart from a few exceptions, the maximum permissible quantities are between 0.15 and 0.25%.

21.9
Antimicrobial Action

21.9.1
General Criteria of Action

The antimicrobial action of benzoic acid is based on various interventions in the enzymatic structure of the cell of the microorganism. In many bacteria and yeasts, for instance, enzymes that control the acetic acid metabolism and oxidative phosphorylation are inhibited (Bosund 1960, Bosund 1962). Benzoic acid appears to intervene at various points in the citric acid cycle, especially that of α-ketoglutaric acid and succinic acid dehydrogenase (Bosund 1962). Benzoic acid also appears to inhibit tyrosinase (Menon et al. 1990). Besides its enzyme-inactivating effects benzoic acid also acts on the cell wall.

To be able to develop its action inside the cell of the microorganism, benzoic acid has to penetrate the cell wall. When this takes place, it is the undissociated part of the acid that more readily enters the cell. Hence, the action of the benzoic acid is dependent on the pH value. Only the undissociated part of the acid has an antimicrobial action. Owing to its relatively high dissociation constant of $6.46 \cdot 10^{-5}$, benzoic acid can be used only for preserving strongly acid products. Depending on the concentration, benzoic acid lowers the intracellular pH value as well. This effect, which likewise helps to inhibit growth and destroy the cells (Salmond et al. 1984), is more pronounced in the undissociated acid portion (Eklund 1985).

No resistance in the true sense of the term occurs, i.e. there is no increase in the minimum inhibitory concentration under the influence of sub-threshold benzoic acid concentrations (Lück and Rickerl 1959).

21.9.2
Spectrum of Action

On the whole, the action of benzoic acid is directly mainly against yeasts and molds, including aflatoxin-forming microorganisms (Uraih and Chipley 1976, Uraih et al. 1977). Bacteria are only partially inhibited. Benzoic acid has less effect on lactic acid bacteria and clostridia. Minimum inhibitory concentrations of benzoic acid for use against some of the bacteria, yeasts and molds important in food spoilage are given in Tables 28 to 30 (Rehm 1961, Balatsouras and Polymenacos 1963).

21.10
Fields of Use

21.10.1
Fat Products

For a number of decades benzoic acid was the favored preservative for margarine, the applied concentration being 0.08 to 0.15%. Benzoic acid is added to the fat

Table 28. Inhibitory action of benzoic acid on bacteria (Rehm 1961)

Name of test organism	pH value	Minimum inhibitory concentration in ppm
Pseudomonas species	6.0	2000– 4800
Micrococcus species	5.5–5.6	500– 1000
Streptococcus species	5.2–5.6	2000– 4000
Lactobacillus species	4.3–6.0	3000–18000
Escherichia coli	5.2–5.6	500– 1200
Bacillus cereus	6.3	5000

Table 29. Inhibitory action of benzoic acid on yeasts (Balatsouras and Polymenacos 1963)

Name of test organism	pH value	Minimum inhibitory concentration in ppm
Sporogenic yeasts	2.6–4.5	200–2000
Asporogenic yeasts	4.0–5.0	700–1500
Hansenula subpelliculosa		2000–3000
Pichia membranaefaciens		7000
Pichia pastori		3000
Candida krusei		3000–7000
Torulopsis species		2000–5000
Rhodotorula species		1000–2000
Oospora lactis		3000

Table 30. Inhibitory action of benzoic action on molds (Rehm 1961)

Name of test organism	pH value	Minimum inhibitory concentration in ppm
Rhizopus nigricans	5.0	300–1200
Mucor racemosus	5.0	300–1200
Penicillium species	2.6–5.0	300–2800
Penicillium glaucum	5.0	4000–5000
Aspergillus species	3.0–5.0	200–3000
Cladosporium herbarum	5.1	1000

phase and/or sodium benzoate to the water phase in appropriate quantities. Actually, benzoic acid is not an ideal preservative for margarine because the relatively high pH value of the margarine is at the extremity of the range in which benzoic acid produces its optimum effect. Consequently, and also because the distribution coefficient between the fat and water phases is not very favorable, benzoic acid is now of only minor importance for margarine preservation.

Benzoic acid and sodium benzoate remain important in the preservation of mayonnaise and mayonnaise-containing delicatessen products, which, as oil-in-water emulsions, are more susceptible to microbial attack than emulsions of the water-in-oil type. In this instance sodium benzoate is usually employed in combination with potassium sorbate, this mixture having a better effect against acid producing bacteria than the two components individually. Moreover, the potassium sorbate content renders the mixture less obtrusive organoleptically than sodium benzoate on its own.

21.10.2 Fruit and Vegetable Products

Benzoic acid used to be employed on a large scale in the form of 0.1–0.2% sodium benzoate for preserving pickled vegetables, an application for which it is well suited owing to the low pH of these products. Moreover, the high acid content of the pickled vegetables, together with the spices employed, restricts the adverse effect of the benzoate on the flavor. Because of changes in eating habits, this field of use has diminished greatly in importance in favor of pasteurization and sorbic acid.

In principle, benzoic acid is a good preservative for acid fruit products; in fact, it is still used for these because of its low price, even though the risk of impairing the flavor is substantially greater than with sorbic acid. For reasons of solubility, benzoic acid is used in this application mainly in the form of sodium benzoate. Fruit pulps can be protected by additions of 0.1 to 0.13% sodium benzoate against mold attack and fermentation. Like sorbic acid, benzoic acid is ineffective against oxidation and enzymatic spoilage, and is therefore as a general rule combined with small quantities of sulfur dioxide or other antioxidants when used in commercial applications for fruit products.

21.10.3 Drinks

For the preservation of pure fruit juices, the remarks stated in conjunction with fruit pulps apply. Benzoic acid is used in the form of sodium benzoate mainly to preserve fruit juices intended for further processing. In general, sodium benzoate is combined with small quantities of SO_2 in order to protect the products against oxidation, enzymatic spoilage and bacterial spoilage (lactic acid and acetic acid fermentation). In addition, fruit juices are pasteurized to inactivate the enzymes and reduce the microbe count. The applied concentration of sodium benzoate is 0.05–0.2%, depending on the types of juices and the length of time for which it is desired that the products should be kept fresh. In the case of soft drinks, sodium benzoate in a dosage of 0.02% provides an inexpensive additional protection against spoilage by yeasts. In soft drinks, benzoates in the presence of ascorbic acid and metal ions can form benzene in the ppb range (Page et al. 1992, McNeil 1993, Gardner and Lawrence 1993).

21.11 Other Effects

Benzoic acid and its salts may be readily perceptible in the flavor of foods they are used to preserve, even in the dosages required for preservation.

21.12 Literature

Balatsouras GD, Polymenacos NG (1963) Chemical preservatives as inhibitors of yeast growth. J Food Sci 28, 267–275

Bedford PGC, Clarke EGC (1972) Experimental benzoic acid poisoning in the cat. Vet Rec 90, 53–58

Bosund I (1960) The bacteriostatic action of benzoic and salicylic acids. II. The effect on acetate metabolism. Acta Chem Scand 14, 111–125

Bosund I (1962) The action of benzoic and salicylic acids on the metabolism of microorganisms. Adv Food Res 11, 331–353

Deuel HJ, Alfin-Slater R, Weil CS, Smyth HF (1954) Sorbic acid as a fungistatic agent for foods. I. Harmlessness of sorbic acid as a dietary component. Food Res 19, 1–12

Eklund T (1985) Inhibition of microbial growth at different pH levels by benzoic and propionic acids and esters of *p*-hydroxybenzoic acid. Int J Food Microbiol 2, 159–167

Federation of American Societies for Experimental Biology (1973) Evaluation of the health aspects of benzoic acid and sodium benzoate as food ingredients. PB-223 837. National Technical Information Service. US Department of Commerce, Springfield

Food and Drug Research Laboratories, Inc (1972) Teratologic evaluation of FDA 71–37 (sodium benzoate). PB-221 777. National Technical Information Service. US Department of Commerce, Springfield

Fujitani T (1993) Short-term effect of sodium benzoate in F 344 rats and B6C3F1 mice. Toxicol Letters 69, 171–179

Gardner L, Lawrence G (1993) Benzene production from decarboxylation of benzoic acid in the presence of ascorbic acid and a transition-metal catalyst. J Agric Food Chem 41, 693–695

Hagenauer-Hener U, Frank C, Hener U, Mosandl A (1990) Bestimmung von Aspartam, Acesulfam-K, Saccharin, Coffein, Sorbinsäure und Benzoesäure in Lebensmitteln mittels HPLC. Dtsch Lebensm-Rundsch 86, 348–351
Informatics, Inc (1972) GRAS (Generally Recognized as Safe) food ingredients: Benzoic acid and sodium benzoate. PB-221 208. National Technical Information Service. US Department of Commerce, Springfield
Jacobsen D (1992) Adverse reactions to benzoates and parabens. In: Food Allergy. Blackwell, Oxford, p. 276–287
JECFA (1993) WHO Techn Rep Series 837, p. 36. WHO: Geneva
Kahn S, Murawski M, Sherma J (1994) Quantitative high performance thin layer chromatographic determination of organic preservatives in beverages. J Liquid Chromatogr 17, 855–865
Kieckebusch W, Lang K (1960) Die Verträglichkeit der Benzoesäure im chronischen Fütterungsversuch. Arzneim Forsch 10, 1001–1003
Kreis H, Frese K, Wilmes G (1967) Physiologische und morphologische Veränderungen an Ratten nach peroraler Verabreicherung von Benzoesäure. Food Cosmet Toxicol 5, 505–511
Lang K (1960) Verträglichkeit der Benzoesäure. Z Lebensm Unters Forsch 112, 394–403
Lück H, Rickerl E (1959) Untersuchungen an *Escherichia coli* über eine Resistenzsteigerung gegen Konservierungsmittel und Antibiotica. Z Lebensm Unters Forsch 109, 322–329
Marquardt P (1960) Zur Verträglichkeit der Benzoesäure. Arzneim Forsch 10, 1033
McNeil T, Nyman J, Diachenko W, Hollifield C (1993) Survey of benzene in foods by using headspace concentration techniques and capillary gas chromatography. J Assoc Off Anal Chem Int 76, 1213–1219
Menon S, Fleck R, Yong G, Strothkamp K (1990) Benzoic acid inhibition of α, β and γ-isoenzymes of *Agaricus bisporus tyrosinase*. Arch Biochem Biophys 280, 27–32
Michils A, Vandermoten G, Duchateau J, Yernault J-C (1991) Anaphylaxis with sodium benzoate. Lancet 337, 1414–1425
Minor JL, Becker BA (1971) A comparison of the teratogenic properties of sodium salicylate, sodium benzoate and phenol. Toxicol Appl Pharmcol 19, 373
Moors M, Teixera C, Jimidar M, Massart D (1991) Solid-phase extraction of the preservatives sorbic acid and benzoic acid and the artificial sweeteners aspartame and saccharin. Anal Chim Acta 255, 177–186
O'Connor J, Costell M, Grisolia S (1989) Carbamyl-glutamate prevents the potentiation of ammonia toxicity by sodium benzoate. Eur J Pediatry 148, 540–542
Page D, Conacher H, Weber D, Lacroix G (1992) A survey of benzene in fruits and retail fruit juices, fruit drinks and soft drinks. J Assoc Off Analy Chem Int 75, 334–340
Rehm H-J (1961) Grenzhemmkonzentrationen der zugelassenen Konservierungsmittel gegen Mikroorganismen. Z Lebensm Unters Forsch 115, 293–309
Reifschneider C, Klug C, Jager M (1994) Konservierungsstoffe in kosmetischen Mitteln. Identifizierung und Quantifizierung. SÖFW-Journal 120, 650–654
Sado I (1973) Synergistic toxicity of official permissible preservative food additives. Nippon Eiseigaku Zasshi 28, 463–476
Salmond CV, Kroll RG, Booth IR (1984) The effect of food preservatives on pH homeostasis in *Escherichia coli*. J Gen Microbiol 130, 2845–2850
SCF (1994) Report on the 92nd session. Brussels
Schaubschläger W, Becker W, Schade U, Zabel P, Schlaak M (1991) Release of mediators from human gastric mucosa and blood in adverse reactions to benzoate. Int Arch Allergy Appl Immunol 96, 97–101
Seiber R, Bütikofer U, Baumann E, Bosset J (1990) Über das Vorkommen der Benzoesäure in Sauermilchprodukten und Käse. Mitt Geb Lebensmittelunters Hyg 81, 484–493
Shtenberg AJ, Ignat'ev AD (1970) Toxicological evaluation of some combinations of food preservatives. Food Cosmet Toxicol 8, 369–380
Strahlmann B (1974) Entdeckungsgeschichte antimikrobieller Konservierungsstoffe für Lebensmittel. Mitt Geb Lebensmittelunters Hyg 65, 96–130

Tremblay G, Aureshi I (1993) The biochemistry and toxicology of benzoic acid metabolism and its relationship to the elimination of waste nitrogen. Pharmac Ther 60, 63–90
Uraih N, Chipley JR (1976) Effects of various acids and salts on growth and aflatoxin production by *Aspergillus flavus* NRRL 3145. Microbios 17: 67, 51–59
Uraih N, Cassity TR, Chipley JR (1977) Partial characterization of the mode of action of benzoic acid on aflatoxin biosynthesis. Can J Microbiol 23, 1580–1584
Waldo JF, Masson JM, Lu WC, Tollstrup J (1949) The effect of benzoic acid and caronamide on blood penicillin levels and on renal function. Am J Med Sci 217, 563–568

Chapter 22

Esters of *p*-Hydroxybenzoic Acid

22.1 Synonyms

IUPAC: Esters of 4-Hydroxybenzoic Acid
English: Parabens. *German:* Ester der *p*-hydroxybenzoesäure, *p*-hydroxybenzoesäureester, PHB-Ester, Solbrol, Nipaester, Nipagin, Nipasol, Nipakombin. *French:* Esters de l'acide *p*-hydroxybenzoique. *Italian:* Esteri *p*-idrossibenzoici. *Spanish:* Esteres del ácido *p*-hidroxibenzoico. *Russian:* Эфиры р-оксибенэойной кислоты.

22.2 History

The esters of *p*-hydroxybenzoic acid were produced with a view to discovering a replacement for salicylic acid and benzoic acid, both of which suffer from the drawback of being effective only in the highly acid pH range. Presuming that esters might be advantageous in this respect, T. Sabalitschka synthesized various alkyl and aryl esters of *p*-hydroxybenzoic acid in the early 1920s, and in 1923 proposed these as preservatives for foods and pharmaceuticals (Strahlmann 1974). Despite certain favorable technical properties and better toxicological characteristics than preservatives familiar up to then, the *p*-hydroxybenzoic acid esters established themselves on only a very limited scale in food preservation because of their organoleptic properties. Currently they are used chiefly in the preservation of pharmaceutical and cosmetic preparations, this being a field in which the main advantage of these products can be utilized, namely continued efficacy even in a weakly acid to neutral pH range.

22.3 Properties

p-Hydroxybenzoic acid methyl ester, molar mass 152.15, water solubility at room temperature 0.25 g/100 g, *p*-hydroxybenzoic acid ethyl ester, molar mass 166.18, water solubility 0.17 g/100 g, *p*-hydroxybenzoic acid propyl ester, molar mass 180.20, water solubility 0.02 g/100 g, and *p*-hydroxybenzoic acid *n*-heptyl ester, molar mass 236.21, water solubility 0.0015 g/100 g, white crystalline powders with a slightly anesthetic taste and melting points between 127°C (methyl ester) and 97°C (propyl ester). The sodium salts of *p*-hydroxybenzoic acid esters are readily water soluble but their storage stability is not very good since hydrolysis occurs rapidly as a result of their strong alkalinity. The solubilities of the *p*-hydroxybenzoic acid esters in fatty oils rise with an increase in the chain length of the alcohol component and vary between 2 and 5%, depending on the nature of the oil. Owing to their high oil-solubility, these products have relatively unfavorable distribution coefficients between the oil and water phases of emulsions (Lubieniecki-von Schelhorn 1967 a and b).

OH

COO-R

$R = CH_3, C_2H_5, C_3H_7$

22.4 Analysis

The esters of *p*-hydroxybenzoic acid can be isolated from the foods to be investigated by means of ether, ether/petroleum ether mixtures or acid steam distillation. After being saponified, these esters can be determined spectrophotometrically in the distillate at 255 nm as *p*-hydroxybenzoic acid (Lorenzen and Sieh 1962). The established methods for identification and quantitative determination of *p*-hydroxybenzoic acid esters are HPLC and TLC (Flak and Schaber 1988, Olea Serrano et al. 1991, Reifschneider et al. 1994).

22.5 Production

The esters of *p*-hydroxybenzoic acid are obtained by reacting the corresponding alcohols with *p*-hydroxybenzoic acid, which can itself be obtained by means of a modified Kolbe reaction from dry potassium phenolate and carbon dioxide under pressure at 180 to 250 °C.

22.6 Health Aspects

22.6.1 Acute Toxicity

The acute toxicity of *p*-hydroxybenzoic acid esters with low-molecular alcohols declines with a reduction in the chain length of the alcohol component (Informatics 1972). The LD_{50} of the methyl and propyl esters after being fed to mice is in the region of 8 g/kg body weight (Sokol 1952, Matthews et al. 1956); whilst the respective figures for the sodium compounds are 2 g and 3.7 g/kg body weight (Matthews et al. 1956). The LD_{50} of the sodium compounds of the ethyl and butyl esters has been determined as 2 g and 0.95 g/kg body weight respectively (Matthews et al. 1956). According to other sources, the LD_{50} of the ethyl ester is 6 g, that of the propyl ester 6.3 g and that of the butyl ester 13.2 g/kg body weight (Sado 1973). For rabbits the LD_{50} for methyl ester is some 3 g/kg body weight (Schübel and Manger 1929). The acute toxicity of *p*-hydroxybenzoic acid esters is not increased by their combination with other preservatives (Sado 1973).

22.6.2 Subchronic Toxicity

A dosage of 500 mg *p*-hydroxybenzoic acid methyl ester per kg body weight per day caused no damage to rabbits over a 6-day period although 3000 mg did prove toxic. Cats and dogs are more sensitive. The subchronic toxicity for propyl ester is of a similar order (Schübel and Manger 1929). Doses of 2 g methyl and propyl ester per day were tolerated by humans over one month without damage (Sabalitschka and Neufeld-Crzellitzer 1954). After sub-chronic oral administration to male and female rats of 2.5 and 5% isopropyl *p*-hydroxybenzoic acid ester respectively, a significant reduction in body weight of rats was observed (Onadera et al. 1994).

22.6.3 Chronic Toxicity

In a long-term feeding trial lasting for virtually two years, 0.9 to 1.2 g *p*-hydroxybenzoic acid methyl, ethyl, propyl and butyl esters per kg body weight caused no damage attributable to these substances specifically, nor were any histological changes to the kidneys, liver, lungs or other internal organs observed. In young rats, 5.5 to 5.9 g/kg body weight daily caused growth disorders. Dogs tolerate 1 g/kg body weight daily over one year at least (Matthews et al. 1956).

The SCF has set a temporary ADI value of 0–10 mg/kg, after cell proliferation studies on the forestomach of rats produced positive results (SCF 1994).

p-Hydroxybenzoic acid methyl ester is not carcinogenic (Informatics 1972).

22.6.4 Intolerance Reactions

Like benzoic acid and certain azo dyes, *p*-hydroxybenzoic acid esters are food additives with considerable sensitizing potential. After both oral administration and topical application *p*-hydroxybenzoic acid esters may cause intolerance reactions, such as urticaria. These are mostly pseudo-allergic reactions, but genuine allergies have also been reported (Jacobsen 1992).

22.6.5 Biochemical Behavior

p-Hydroxybenzoic acid esters are rapidly and completely absorbed from the gastro-intestinal tract and hydrolyzed (Matthews et al. 1956). Consequently no accumulation is likely to occur. The *p*-hydroxybenzoic acid formed in the hydrolysis is excreted via the urine. In the case of the rat, this occurs within 24 hours, some 40% being excreted as such, 23% as *p*-hydroxyhippuric acid and a further 23% as esters of glucuronic acid. Only small concentrations of *p*-hydroxybenzoic acid are detectable in the blood (Derache and Gourdon 1963). In the rabbit, the pattern of excretion is basically the same and only the quantity of the metabolites is somewhat different (Tsukamoto and Terada 1960, 1962, 1964). Excretion in cats (Schübel and Manger 1929) and dogs (Matthews et al. 1956) is also similar. After peroral administration of 2 g propyl ester to humans over several days, the urine contains some 14% of the administered ester quantity in the form of *p*-hydroxybenzoic acid, 4% in the form of *p*-hydroxyhippuric acid and the remainder probably combined with sulfuric acid (Sabalitschka and Neufeld-Crzellitzer 1954). Methyl, ethyl, propyl and butyl esters have a spermicidal effect on human sperm. Their use in contraceptives has therefore been discussed (Bao-Liang et al. 1989).

22.7 Regulatory Status

The methyl, ethyl and *n*-propyl esters of *p*-hydroxybenzoic acid, as well as their sodium compounds, are permitted in the majority of countries for preservation of many foods. The maximum permissible quantity, other than in exceptional cases, is between 0.1 and 0.2%.

22.8 Antimicrobial Action

22.8.1 General Criteria of Action

The antimicrobial action of the *p*-hydroxybenzoic acid esters is proportional to the chain length of the alcohol component (Aalto et al. 1953, Shibasaki 1969, Thompson 1994). Thus the antimicrobial action of the methyl ester is some 3 to 4 times, that of the ethyl ester some 5 to 8 times, that of the propyl ester some 17 to 25 times, that of the butyl ester some 30 to 40 times, and that of the benzyl ester some 70 to 110 times as powerful as phenol. However, superimposed on this effect may be the water solubility of the esters, which is inversely proportional to the chain length of the alcohol component. In addition, as in the case of phenol, *p*-hydroxybenzoic acid esters may be linked to some extent to proteins, emulsifiers and other substrate constituents on account of their phenolic OH group and thus be inactivated. In their mechanism of action, the *p*-hydroxybenzoic acid esters correspond in principle to phenol. One of their actions involves the destruction of the cell membrane and protein denaturation in the interior of the cell, besides which competitive reactions occur with coenzymes.

As *p*-hydroxybenzoic acid esters are non-dissociating compounds, their antimicrobial action is relatively independent of the pH value of the medium to be preserved. In this respect they are superior to the organic preservative acids.

Resistance in the true sense of the term, i.e. an increase in the minimum inhibitory concentrations under the influence of sub-threshold concentrations of *p*-hydroxybenzoic acid esters, does not occur (Lück and Rickerl 1959).

The antimicrobial action is based on inhibition of the absorption of essential nutrients and amino acids, such as glucose and proline, and on destruction of the complex structure of the cell membrane (Tatsuguchi et al. 1991a and 1991b).

22.8.2 Spectrum of Action

Like benzoic acid and sorbic acid, the *p*-hydroxybenzoic acid esters belong to the preservatives with a primarily fungistatic action. Nevertheless, the bacteriostatic action of the esters is somewhat higher than that of the two preservative acids mentioned, owing to the phenolic OH group of the former. This applies especially to the gram-positive strains. Since the action of the individual *p*-hydroxybenzoic acid esters varies somewhat, they are occasionally employed in mixtures. Minimum inhibitory concentrations of the *p*-hydroxybenzoic acid esters for some of the bacteria, yeasts and molds involved in food spoilage are given in Tables 31–33 (Rehm 1961).

Table 31. Inhibitory action of *p*-hydroxybenzoic acid esters on bacteria at pH 5.5 to 7.0 (Rehm 1961)

Name of test organism	Minimum inhibitory concentration in ppm	
	Ethyl ester	Propyl ester
Pseudomonas species	500–1000	400–1000
Micrococcus species	600–1100	100–1000
Streptococcus faecalis	1300	400
Lactobacillus species	350–1500	300– 500
Betabacterium buchneri	4000	4000
Escherichia coli	120–1000	300–1000
Bacillus cereus	1000	100–1000
Salmonella species		330– 770

Table 32. Inhibitory action of *p*-hydroxybenzoic acid esters on yeasts at pH 3.0–5.0 (Rehm 1961)

Name of test organism	Minimum inhibitory concentration in ppm	
	Ethyl ester	Propyl ester
Sporogenic yeasts	500–1000	150–600
Saccharomyces cerevisiae	800	400
Asporogenic yeasts	200– 800	200–400
Candida species	200– 800	500
Torula lipolytica	600	300

Table 33. Inhibitory action of *p*-hydroxybenzoic acid esters on molds at pH 5–7 (Rehm 1959, 1961)

Name of test organism	Minimum inhibitory concentration in ppm	
	Ethyl ester	Propyl ester
Phycomycetae	200–1000	200–1000
Mucor racemosus	200– 500	100– 500
Rhizopus nigricans	200– 500	200– 500
Penicillium species	200– 800	200– 500
Gliocladium roseum	600	300
Aspergillus species	200– 500	100– 500
Aspergillus niger	500–1000	100– 500
Aspergillus orycae	200– 500	100– 200
Fungi imperfecti	200– 400	200– 300

22.9 Fields of Use

The uses of *p*-hydroxybenzoic acid esters in food preservation are governed by three properties, namely an action independent of the pH value, very low water solubility and particular organoleptic characteristics.

At one time the good action of *p*-hydroxybenzoic acid at high pH values was utilized in preserving the surfaces of hard cheese and dried sausage against mold attack – applications in which the unfavorable sensory characteristics presented little or no problem. Like the preservation of confectionery and fish products, however, these fields of use are ones where these preservatives are no longer of any great importance.

Although the use of *p*-hydroxybenzoic acid esters derived from low-molecular alcohols is of no importance in drinks preservation, *p*-hydroxybenzoic acid *n*-heptyl ester is used in the USA on a small scale in the preservation of beer, which it protects against yeasts and bacterially induced post-fermentation after brewing is completed. The applied concentration in this case is 8–12 g/1000 l (Kozulis et al. 1971), which is approximately the same level as the solubility. To incorporate the *p*-hydroxybenzoic acid ester into beer, the preservative is dissolved in alkali before, or in propylene glycol. In its usual concentrations, *p*-hydroxybenzoic acid *n*-heptyl ester reduces the stability of the froth and the low-temperature stability of the beer itself.

22.10 Other Effects

The esters of *p*-hydroxybenzoic acid are readily perceptible in the taste of the products they are used to preserve, even in the dosages required for preservation.

In concentrations of over 0.1%, *p*-hydroxybenzoic acid esters can act as local anesthetics and spasmolytics.

22.11 Literature

Aalto TR, Firman MC, Rigler NE (1953) *p*-hydroxybenzoic acid esters as preservatives. I. Uses, antibacterial and antifungal studies, properties and determination. J Am Pharma Assoc Sci Ed 42, 449–457

Bao-Liang S, Hai-Ying L, Dun-Ren P (1989) In vitro spermicidal activity of parabens against human spermatozoa. Contraception 39, 331–335

Derache R, Gourdon J (1963) Métabolisme d'un conservateur alimentaire: L'acide parahydroybenzoique et ses esters. Food Cosmet Toxicol 1, 189–195

Davidson M (1993) Parabens and phenolic compounds. In: Davidson M, Branen AL (eds) Antimicrobials in foods. Marcel Dekker, New York, p. 263–306

Flak W, Schaber R (1988) Die Bestimmung von Konservierungsmitteln in Weinen und anderen Getränken mittels Hochdruckflüssigkeitschromatographie. Mitt Klosterneuburg 38, 10–16

Informatics Inc (1972) GRAS (Generally Recognized as Safe) Food Ingredients: Methyl and propyl paraben. PB-221 209. National Technical Information Service. US Department of Commerce, Springfield

Jacobsen D (1992) Adverse reactions to benzoates and parabens. In: Food allergy. Blackwell, Boston, p. 276–287

Kozulis JA, Bayne PD, Cuzner J (1971) New technique for the cold sterilization of beer. Ann Proc Am Soc Brewing Chem 105–115

Lorenzen W, Sieh R (1962) Spektrophotometrische Schnellbestimmung von Konservierungsstoffen in Lebensmitteln. Z Lebensm Unters Forsch 118, 222–233

Lubieniecki-von Schelhorn M (1967 a) Untersuchungen über die Verteilung von Konservierungsstoffen zwischen Fett und Wasser. I. Mitteilung. Physikalisch-chemische Untersuchungen. Z Lebensm Unters Forsch 131, 329–345

Lubieniecki-von Schelhorn M (1967 b) Untersuchungen über die Verteilung von Konservierungsstoffen zwischen Fett und Wasser. II. Mitteilung. Beziehungen zwischen physikalisch-chemischer Verteilung und antimikrobieller Wirksamkeit von Konservierungsstoffen in fetthaltigen Lebensmitteln. Z Lebensm Unters Forsch 133, 227–241

Lück H, Rickerl E (1959) Untersuchungen an *Escherichia coli* über eine Resistenzsteigerung gegen Konservierungsmittel und Antibiotica. Z Lebens Unters Forsch 109, 322–329

Matthews C, Davidson J, Bauer E, Morrison JL, Richardson AP (1956) *p*-hydroxybenzoic acid esters as preservatives. II. Acute and chronic toxicity in dogs, rats and mice. J Am Pharm Assoc Sci Ed 45, 260–267

Olea Serrano F, Sanchez I, Rivilla G (1991) High performance liquid chromatography determination of chemical preservatives in yogurt. J Liquid Chromatogr 14, 709–717

Onadera H, Mitsumori K, Yasuhara K, Shimo T, Kurokawa N, Takahashi M (1994) 13-week subchronic oral toxicity study of isopropyl-*p*-hydroxybenzoate in F 344 rats. Bull Natl Inst Health Sci Japan 112, 82–88

Rehm H-J (1959) Erfahrungen mit der Filterpapiermethode zur Schnellprüfung von Konservierungsmitteln gegen Schimmelpilze. Z Lebensm Unters Forsch 110, 375–381H-J (1961) Grenzhemmkonzentrationen der zugelassenen Konservierungsmittel gegen Mikroorganismen. Z Lebensm Unters Forsch 115, 293–309

Reifschneider C, Klug C, Jager M (1994) Konservierungsstoffe in kosmetischen Mitteln – Identifizierung und Quantifizierung. SÖFW-J 120, 650–654

Sabalitschka T, Neufeld-Crzellitzer R (1954) Zum Verhalten der *p*-Oxybenzoesäureester im menschlichen Körper. Arzneim Forsch 4, 575–579

Sado I (1973) Synergistic toxicity of official permissible preservative food additives. Nippon Eiseigaku Zasshi 28, 463–476

SCF (1994) Report on the 92nd session. Brussels
Schübel K, Manger J (1929) Ein Beitrag zur Pharmakologie einiger Paraoxybenzoesäureester: das Schicksal im Organismus und die Toxizität. Naunyn Schmiedebergs Arch Exp Pathol Pharmakol 146, 208–222
Shibasaki I (1969) Antimicrobial activity of alkyl esters of *p*-hydroxybenzoic acid. J Ferment Technol 47, 167–177
Sokol H (1952) Recent developments in the preservation of pharmaceuticals. Drug Stand 20, 89–106
Strahlmann B (1974) Entdeckungsgeschichte antimikrobieller Konservierungsstoffe für Lebensmittel. Mitt Geb Lebensmittelunters Hyg 65, 96–130
Tatsuguchi K, Kuwamoto S, Ogomori M, Ide T, Watanabe T (1991 a) Membrane disorders of *Escherichia coli* cells and liposomes induced by *p*-hydroxybenzoic acid esters. J Food Hyg Soc Japan 32, 121–127
Tatsuguchi K, Kuwamoto S, Watanabe T (1991 b) Membrane degradation of heat-injured *Escherichia coli* stimulated by *p*-hydroxybenzoate. J Food Hyg Soc Japan 32, 278–283
Thompson DP (1994) Minimum inhibitory concentration of esters of *p*-hydroxybenzoic acid (paraben) combinations against toxigenic fungi. J Food Protect 57, 133–135
Tsukamoto H, Terada S (1960) Metabolism of drugs. XXIII. Metabolic fate of *p*-hydroxybenzoic acid and its derivatives in rabbit. Chem Pharm Bull 8, 1066–1070
Tsukamoto H, Terada S (1962) Metabolism of drugs. XXVI. Metabolic fate of *p*-hydroxybenzoic acid and its derivatives in rabbit. Chem Pharm Bull 10, 86–90
Tsukamoto H, Terada S (1964) Metabolism of drugs. XXVII. Metabolic fate of *p*-hydroxybenzoic acid and its derivatives in rabbit. Chem Pharm Bull 12, 765–769

o-Phenylphenol

23.1 Synonyms

IUPAC: Biphenyl-2-ol
English: *o*-Hydroxybiphenyl, Dowicide 1, orthoxenol, SOPP. *German:* *o*-Phenylphenol, 2-Phenylphenol, *o*-Oxydiphenyl, *o*-Hydroxybiphenyl, Orthophenylphenol, Preventol O extra. French: *o*-Phenylphenol. *Italian:* *o*-Fenilfenolo. *Spanish:* *o*-Fenilfenol. *Russian:* о-фенилфенол.

23.2 History

The antimicrobial action of the phenol derivative, *o*-phenylphenol, has been known for a long time, the product having first been suggested as a mold preventive for citrus fruit in Britain in the 1930s. Like biphenyl, *o*-phenylphenol was adopted for citrus fruit preservation in the postwar years.

23.3 Properties

$C_{12}H_{10}O$, molar mass 170.21, white to yellowish-white flakes that melt at 57 °C. *o*-Phenylphenol is only very sparingly soluble in water, whereas the sodium compound, white flakes or white powder, is very readily water-soluble. At room temperature, some 120 g sodium-*o*-phenylphenolate dissolve in 100 g water. Trade names employed for sodium-*o*-phenylphenolate are Preventol ON extra and Dowicide A.

OH

23.4 Analysis

For analytical detection, *o*-phenylphenol is isolated by steam distillation from the food to be investigated. After coupling with Fast Blue Salt B, *o*-phenylphenol yields a color with an absorption maximum at 480 nm for photometric determination

(Böhme and Hofmann 1961). *o*-Phenylphenol can also be determined by HPLC (Apelbaum and Barkain-Golan 1987, Luckas 1987).

23.5 Health Aspects

The LD_{50} of *o*-phenylphenol has been determined as 0.5 g/kg body weight for the cat and 3 g/kg body weight for the rat (Macintosh 1945, Hodge et al. 1952).

Over a period of 32 days up to 200 mg *o*-phenylphenol per kg body weight per day administered to the rat produced no apparent damage to health (Macintosh 1945).

In a two-year feeding experiment 0.2% *o*-phenylphenol in the feed of rats had no influence on the growth rate, mortality, appearance or behavior of the animals. The histological condition of various internal organs corresponds to those of the control animals. 2% *o*-phenylphenol led to reduced weight increase and renal damage. Dogs tolerate 0.5 g/kg body weight over a period of 1 year without incurring damage due specifically to the product in question (Hodge et al. 1952).

In less recent literature the carcinogenicity of *o*-phenylphenol is disputed (Hiraga and Fujii 1984, Hagiwara et al. 1984). More recent studies (Fujii et al. 1986 and 1987) indicate carcinogenic effects on the bladder in various experimental animal species and further toxic effects on the liver, kidneys and bladder. *o*-Phenylphenol has not been proved to have a carcinogenic effect on human subjects. Nevertheless it is classified by the MAK committee in list IIb (substances for which a maximum allowable workplace concentration has not yet been defined) and is regarded as mutagenic. It appears that the effects are caused not by *o*-phenylphenol itself, but by its metabolites phenylhydroquinone and phenylbenzoquinone (Kolachana et al. 1991). Corresponding DNA adducts have been detected both in vitro and in vivo (Ushiyama et al. 1992).

Orally administered *o*-phenylphenol is partly converted by the rat into 2.5-dihydroxybiphenyl, which, like *o*-phenylphenol itself, is excreted in the urine as glucuronic acid or sulfuric acid conjugate (Souci et al. 1967).

23.6 Regulatory Status

o-Phenylphenol is permitted in most countries for the surface treatment of citrus fruit. The maximum permissible residual quantity is 12 mg/kg fruit.

23.7 Antimicrobial Action

Like biphenyl, *o*-phenylphenol in concentrations of more than 5 ppm inhibits the carotin synthesis of microorganisms. In other respects, *o*-phenylphenol corresponds to other phenol derivatives in its criteria of antimicrobial action on account of its phenolic OH group. A further aspect of its antimicrobial action is that of its non-specific denaturing effects on the cell wall of microorganisms. In addition, *o*-phenylphenol inhibits various enzyme systems of the cell, e.g. NAD-oxidase (Souci et al. 1967). Owing to the introduction of an additional aromatic nucleus, *o*-phenylphenol has greater lipid solubility than phenol. As a result, *o*-phenylphenol has a more powerful action than phenol. Unlike preservative acids, the antimicrobial action of *o*-phenylphenol increases with a rise in pH value. The dissociated compound has a more powerful action than that of the undissociated compound (Rehm et al. 1966).

Like most phenol derivatives, *o*-phenylphenol has a fairly broad spectrum of action, its effect being both bacteriostatic and fungistatic. So far as food preservation is concerned, however, only its action against molds is of commercial importance. Depending on their type, these are inhibited by concentrations of 10 to 50 ppm in the pH range of 6 to 8. The minimum inhibitory concentration for *Escherichia coli* is somewhat higher, namely 115 ppm (Rehm et al. 1966).

In common with biphenyl, *o*-phenylphenol may induce resistance in molds. Cases of cross-resistance with biphenyl are also known (Souci et al. 1967).

23.8 Fields of Use

The only field of use for *o*-phenylphenol in food technology is in the preservation of citrus fruit. The fruit are immersed for 30 to 60 seconds at 30 to 35 °C in 0.5–2% aqueous solutions of the sodium salt of *o*-phenylphenol. As *o*-phenylphenol is extremely effective in the alkaline pH range, the immersion solution is adjusted to a pH of approximately 11.7. In addition, 1% hexamethylene tetramine (see Sect. 30.10) is added to prevent brown discoloration of the fruit skin (Long and Roberts 1958, Souci et al. 1967). If hard water is used, a small quantity of sodium ethylenediaminetetraacetate is also added to the immersion solution. After being treated the fruit are rinsed off with clear water. A small residue comprising a few milligrams of *o*-phenylphenol per kg fruit nevertheless remains in the skin (Ihloff and Kalitzki 1961). The use of *o*-phenylphenol in a similar way to biphenyl (see Sect. 24.9) for treating packaging materials is of no commercial importance. Recently, increasing resistance of molds on citrus fruits to *o*-phenylphenol has been reported.

23.9 Literature

Apelbaum A, Barkain-Golan R (1987) Effect of altering treatment conditions during post-harvest application of sodium-*o*-phenylphenate on *o*-phenylphenol residues and rind injuries in various citrus fruits. Food Add Contam 4, 317–324

Böhme H, Hofmann G (1961) Über die photometrische Bestimmung von Diphenyl und *o*-Hydroxy-diphenyl in Citrusfruchtschalen. Z Lebensm Unters Forsch 114, 97–105

Fujii T, Hikuriya M, Kamiya N, Hiraga K (1986) Enhancing effect of thiabendazole on urinary bladder carcinogenesis induced by sodium *o*-phenylphenate in F 344 rats. Food Chem Toxicol 24, 207–211

Fujii T, Nakamura K, Hiraga K (1987) Effect of pH on the carcinogenicity of *o*-phenylphenol and sodium *o*-phenylphenate in the rat urinary bladder. Food Chem Toxicol 25, 359–362

Hagiwara A, Shibata M, Hirose M, Fukushima S, Ito N (1984) Long-term toxicity and carcinogenicity study of sodium *o*-phenylphenate in $B6C3F_1$ mice. Food Chem Toxicol 22, 809–814

Hiraga K, Fujii T (1984) Induction of tumors of the urinary bladder in F 344 rats by dietary administration of *o*-phenylphenol. Food Chem Toxicol 22, 865–870

Hodge HC, Maynard EA, Blanchet HJ, Spencer HC, Rowe VK (1952) Toxicological studies of orthophenylphenol (Dowicide 1). J Pharmacol Exp Ther 104, 202–210

Ihloff ML, Kalitzki M (1961) Ueber Konservierungs- und Schönungsmittel sowie Rückstände von Schädlingsbekämpfungsmitteln bei Importobst. Mitt Geb Lebensmittelunters Hyg 52, 327–339

Kolachana P, Subrahmanyam V, Eastmond D, Smith M (1991) Metabolism of phenylhydroquinone by prostaglandin (H) synthese: possible implications in *o*-phenylphenol carcinogenesis. Carcinogenesis 12, 145–149

Long JK, Roberts EA (1958) The phytotoxic and fungicidal effects of sodium *o*-phenylphenate in controlling green mould wastage in oranges. Austr J Agric Res 9, 609–628

Luckas B (1987) Methodik zur gemeinsamen Erfassung fungicider Pflanzenbehandlungsmittel auf Citrusfrüchten und Obst mit Hilfe der HPLC und selektiver Detektion. Z Lebensm Unters Forsch 184, 195–197

Macintosh FC (1945) The toxicity of diphenyl and *o*-phenyl-phenol. Analyst 70, 334–335

Rehm H-J, Laufer-Heydenreich S, Wallnöfer P (1966) Zur Kenntnis der antimikrobiellen Wirkung von Biphenyl und Derivaten des Biphenyls. II. Mitteilung. Physikalische und chemische Grundlagen. Z Lebensm Unters Forsch 135, 117–122

Souci SW, Rehm H-J, Laufer-Heydenreich S, Herbig G (1967) Zur Kenntnis der antimikrobiellen Wirkung von Biphenyl und Derivaten des Biphenyls. I. Mitteilung. Antimikrobielle und stoffwechselphysiologische Wirkung. Z Lebensm Unters Forsch 134, 209–216

Ushiyama K, Nagai F, Nakagawa A, Kano I (1992) DNA-adduct formation by *o*-phenylphenol metabolite in vivo and in vitro. Carcinogenesis 13, 1469–1473

Chapter 24

Biphenyl

24.1 Synonyms

IUPAC: Biphenyl
English: Diphenyl. *German:* Diphenyl, Biphenyl, Phenylbenzol. *French:* Biphényle, diphényle. *Italian:* Difenile. *Spanish:* Difenilo. *Russian:* Дифенил,

24.2 History

The use of biphenyl for the preservation of citrus fruit was first suggested in the late 1930s, although its antimicrobial action had been known earlier. In the postwar period it has established itself on a large scale for this purpose, its use having spread from the USA. In fact, it was only with the introduction of biphenyl that it became feasible for the first time to transport citrus fruit efficiently over long distances. Recently biphenyl has been increasingly replaced by other substances which have less influence on the organoleptic characteristics of the citrus fruit.

24.3 Properties

$C_6H_5C_6H_5$, molar mass 154.21, colorless flakes with an aromatic odor which melt at 69–71 °C. Biphenyl is very sparingly soluble in water but readily soluble in ether, benzene and hydrocarbons. Vapor pressure at 20 °C is $7.3\ 10^{-3}$ mbar, corresponding to a biphenyl concentration of $4.7\ 10^{-2}$ mg/liter air (Rehm et al. 1966).

24.4 Analysis

In the analytical detection process, biphenyl is isolated by steam distillation from the foodstuff to be investigated. Where the food in question consists of citrus fruits, the distilled biphenyl is separated from the accompanying essential oils by passing the distillate through a special distillation attachment, or by purifying the distillate subsequently by column chromatography. Biphenyl can then be determined spectrophotometrically at 250 nm or after nitration as dinitrodiphenyl, which absorbs at 560 nm. It can also be determined by IR spectroscopy (Paseiro Losada et al. 1990) or by HPLC together with other surface treatment agents (Luckas

1987). A very simple method of sample preparation is employed when GC-MSD or GC-FID is used to detect biphenyl in the peel of citrus fruits (Anklam and Müller 1994).

24.5 Production

Biphenyl is contained in the fraction of coal tar which boils at 230–270 °C, from which it can be produced in pure form. It is also formed by pyrogenic dehydrogenation of benzene.

24.6 Health Aspects

24.6.1 Acute Toxicity

The LD_{50} of biphenyl after peroral administration to rats is 3.3 g (Deichmann et al. 1947) and according to other studies approximately 5 g/kg body weight (Pecchiai and Saffiotti 1957). For rabbits the LD_{50} is 2.4 g (Deichmann et al. 1947) and for cats more than 2.6 g/kg body weight (van Stratum 1964).

24.6.2 Subchronic Toxicity

After 4 weeks, 1% biphenyl in the feed of rats causes growth depression as well as damage to the liver and kidneys (West 1940). Up to 200 mg biphenyl per kg body weight daily had no visibly harmful effects on the health after four weeks (Macintosh 1945), nor were any such effects caused by 300 mg biphenyl per kg body weight daily over 12 days (Rogliani and Procaccini 1956). In a 90-day experiment conducted on rats, 0.5 to 1% in the feed did not impair the health (Ambrose et al. 1960). In humans, doses of 35 mg biphenyl over 13 weeks had no deleterious effects (Souci et al. 1967).

24.6.3 Chronic Toxicity

The ingestion of 50 to 100 mg biphenyl, i.e. 0.25 to 0.5% relative to the feed of rats, produced degenerative changes to the liver and kidneys after two months, although up to the thirteenth month these changes did not intensify (Pecchiai and Saffiotti 1957). In an experiment continued over four generations, in which rats received feed containing 0.01, 0.1 and 1.0% biphenyl, only the highest concentration produced untoward effects. Fertility and the size of litters were markedly reduced in this case, and growth disturbances also occurred (van Stratum 1964). Damage was caused to the health of monkeys fed over a protracted period with 1% biphenyl, relative to the feed (van Stratum 1964).

Biphenyl is not carcinogenic (van Stratum 1964).
The ADI has been set at 0–0.125 mg/kg body weight.

24.6.4
Biochemical Behavior

Rats, rabbits, dogs and pigs convert biphenyl in their metabolism into 4-hydroxybiphenyl and other hydroxybiphenyls and/or diphenylglucuronic acid, all these being substances excreted via the urine (van Stratum 1964, Meyer and Scheline 1976, Meyer et al. 1976 a and 1976 b). Certain metabolites of biphenyl may react with thiol groups of proteins or glutathione (Nakagawa et al. 1993).

24.7
Regulatory Status

Biphenyl is permitted in most countries for the preservation of citrus fruit. The maximum permissible residual quantity is 70 mg/kg fruit.

24.8
Antimicrobial Action

Biphenyl is effective against a large number of microorganisms, this being due partly to its inhibition of carotin synthesis. The only aspect of commercial importance, however, is its action against molds. Owing to the poor solubility of biphenyl in water, it is hard to determine the minimum inhibitory concentration. Biphenyl is known to inhibit the development of *Penicillium italicum* and *Penicillium digitatum* in a concentration of only 0.08 mg/l air, although it has no effect on *Phytophthora citrophthora, Alternaria citri, Sclerotinia sclerotiorum* or *Trichoderma viride* (Souci et al. 1967). Molds may become resistant to biphenyl. Strains whose growth is initially halted by biphenyl vapors may recommence growing after a certain period of time (Souci et al. 1967).

24.9
Fields of Use

The only field of use for biphenyl in food technology is the preservation of citrus fruit. As a general rule the packaging materials (wrapping papers, intermediate layers and inserts for boxes and cartons) are impregnated with biphenyl, the usual applied concentration being 1–5 g biphenyl per m^2. Because of its high vapor pressure, the biphenyl evaporates and fills the gas space between the packaging material and the fruit, some of the biphenyl being absorbed by the skins of the fruit (Tomkins and Isherwood 1945). Only rarely do the residue measurements exceed 50 mg/kg fruit (Ihloff and Kalitzki 1961).

24.10 Other Effects

In the days when relatively high doses of biphenyl were still used, a musty odor was often present on the peel of citrus fruits. Warnings were therefore sometimes issued against using such peel to make other foods.

24.11 Literature

Ambrose AM, Booth AN, DeEds F, Cox AJ (1960) Toxicological study of biphenyl, a citrus fungistat. Food Res 25, 328–336

Anklam E, Müller A (1994) A simple method of sample preparation for analysis of biphenyl residues in citrus fruit peels by gas chromatography. Z Lebensm Unters Forsch 198, 329–330

Deichmann WB, Kitzmiller KV, Dierker M, Witherup S (1947) Observations on the effects of diphenyl, *o*- and *p*-aminodiphenyl, *o*- and *p*-nitrodiphenyl and dihydroxyoctachlorodiphenyl upon experimental animals. J Ind Hyg Toxicol 29, 1–13

Ihloff ML, Kalitzki M (1961) Ueber Konservierungs- und Schönungsmittel sowie Rückstände von Schädlingsbekämpfungsmitteln bei Importobst. Mitt Geb Lebensmittelunters Hyg 52, 327–339

Luckas B (1987) Methodik zur gemeinsamen Erfassung fungicider Pflanzenbehandlungsmittel auf Citrusfrüchten und Obst mit Hilfe der HPLC and selektiver Detektion. Z Lebensm Unters Forsch 184, 195–197

Macintosh FC (1945) The toxicity of diphenyl and *o*-phenyl-phenol. Analyst 70, 334–335

Meyer T, Scheline RR (1976) The metabolism of biphenyl. II. Phenolic metabolites in the rat. Acta Pharm Toxicol 39, 419–432T, Aarbakke J, Scheline RR (1976 a) The metabolism of biphenyl. I. Metabolic disposition of ^{14}C-biphenyl in the rat. Acta Pharm Toxicol 39, 412–418

Meyer T, Larsen JC, Hansen EV, Scheline RR (1976 b) The metabolism of biphenyl. III. Phenolic metabolites in the pig. Acta Pharm Toxicol 39, 433

Nakagawa Y, Sumiko T, Moore G, Moldeus P (1993) Cytotoxic effects of biphenyl and hydroxybiphenyls on isolated rat hepatocytes. Biochem Pharmacol 45, 1959–1965

Paseiro Losada P, Simal Lozano S, Simal Gándara J (1990) Determination of biphenyl residues in citrus fruit by derivative infrared spectrophotometry. J Assoc Off Anal Chem 73, 632–637

Pecchiai L, Saffiotti U (1957) Studio della tossicità del difenile, dell'ossidifenile e della loro miscela ("Dowtherm"). Med Lav 48, 247–254

Rehm H-J, Laufer-Heydenreich S, Wallnöfer P (1966) Zur Kenntnis der antimikrobiellen Wirkung von Biphenyl and Derivaten des Biphenyls. II. Mitteilung. Physikalische und chemische Grundlagen. Z Lebensm Unters Forsch 135, 117–122

Rogliani E, Procaccini S (1956) Assorbimento del difenile de parte di arance e limoni avvolti in carte trattate con difenile. Biochim Appl 3, 193–197

Souci SW, Rehm H-J, Laufer-Heydenreich S, Herbig G (1967) Zur Kenntnis der antimikrobiellen Wirkung von Biphenyl und Derivaten des Biphenyls. I. Mitteilung. Antimikrobielle und stoffwechselphysiologische Wirkung. Z Lebensm Unters Forsch 134, 209–216

Stratum PGC van (1964) Die Toxizität von Diphenyl, einem fungistatischen Schutzmittel für Zitrusfrüchte. Eine Übersicht und Auswertung der Literatur. Bericht Nr. R 1838 des Zentralinstituts für Ernährungsforschung T. N. O. Zeist: T. N. O.

Tomkins RG, Isherwood FA (1945) The absorption of diphenyl and *o*-phenyl-phenol by oranges from treated wraps. Analyst 70, 330–333

West HD (1940) Evidence for the detoxication of diphenyl through a sulfur mechanism. Proc Soc Exp Biol New York 43, 373–375

Chapter 25

Smoke

25.1 Synonyms

German: Räucherrauch, Rauch. *French:* Fumée. *Italian:* Fumo. *Spanish:* Humo. *Russian:* Дым (для копчения).

25.2 History

Smoking, namely the use of smoke from an open fire, is one of the oldest methods of preserving meat. It was known even in prehistoric times and has remained highly important to the present day. The technique of smoking has been considerably improved in the last few years, especially with regard to the method of smoke production.

25.3 Methods of Smoking

Smoke is produced in any incomplete combustion of organic material. The smoke used in the smoking of food is produced by the systematic slow combustion of untreated wood, preferably soft wood from deciduous trees, such as beech, oak, alder and sycamore, or foreign woods like hickory, in the form of small twigs, shavings or sawdust or the cones of coniferous trees (fir or pine cones), often with additions of spices. Coniferous wood is less suitable. Depending on the level of the smoke temperature and the characteristics of the smoke, a distinction is drawn between cold, warm, and hot smoking (Klettner 1979, Tóth 1982). Table 34 provides a survey of the methods.

In cold smoking, cooled smoke or smoke at room temperature is brought into contact with the food to be smoked, either over a period of weeks with low-density smoke (long-duration smoking method) or else for a few days (short-duration smoking method). Cold smoking is the preferred method for precured or dried foods, e.g. raw sausages. In warm smoking the temperature of the smoke is between 25 and 50 °C, whilst that for the hot smoking process as used for parboiled sausages is between 50 and 100 °C. The duration of the hot smoking process is usually 30 minutes to 3 hours. A distinction is drawn between dry and wet smoke according to moisture content (moist or steam smoking). Black smoking is the term used to describe processes with high-soot-content smoke from coniferous wood or resin.

Table 34. Methods of Smoking in Meat Technology

Method	Temperature	Duration	Products
Cold smoking	18 °C (12–24°C)	several days to weeks	raw sausages, raw cured meat, parboiled, sausages, boiled sausages
Moist smoking (rare)	up to 30 °C	2–3 days	rapidly matured raw sausages
Warm smoking (rare)	up to 50 °C	1–3 hours	large (-diameter) parboiled sausage
Hot smoking	60–100 °C	20–60 min.	parboiled sausage, boiled sausage

In former times meat was preserved in the smoke of domestic fires. More recently, however, other more precisely controlled smoking processes have been developed. The main aim of these has been to standardize the smoke as far as possible (Kersken 1973, Kersken 1974 b, Tóth and Potthast 1984). These include the use of slow combustion smoke generators, with electrically heated or gas-heated smoke developers, and friction smoke producers with rotating steel friction discs, the latter being employed mainly in conjunction with hickory wood. In the fluidization method and the steam method of smoke generation, wood shavings are brought into contact with air heated to 200–500 °C and superheated steam respectively, both of which produce smoke. Electrostatic smoking uses an electric field to accelerate the deposition of the smoke particles on the substance being smoked (Kersken 1973).

As alternatives to real smoking processes, increasing use is being made of liquid smoke condensates. These are liquid condensates of freshly produced smoke that have been filtered and clarified until virtually free of tar residues and undesired byproducts, such as aromatic hydrocarbons or nitrosamines. They are employed mainly for flavoring purposes (Tóth 1984), but also have some antimicrobial action (Sofos et al. 1988). They can be used for smoked fish products and other foods (Münkner and Meyer 1994).

25.4 Properties

The smoke used for the smoking of food consists of two components, one comprising particles and the other a gas (Kersken 1973, Tóth 1984). The particle component represents an aerosol of liquid droplets uniformly dispersed in the gas phase and larger particles of fly ash, tar or soot. The gas phase contains hydrocarbons, formaldehyde (Potthast and Eigner 1986) and other aldehydes, low-molecular organic acids, phenols and many other constituents. The total number of substances contained in smoke is estimated to be 5,000 to 10,000. Some 500 individual constituents have so far been detected (Tóth 1982, Tóth 1984, Maga 1988,

Potthast 1988). The preservative effect of the smoke is probably attributable mainly to the phenols (Wittkowski 1985). In the phenolic fraction syringaldehyde predominates in hard wood smoke and 4-methyl guaiakol and guaiakol in soft wood smoke. Some of these compounds determine the flavor of the smoked meat (Maga 1987).

25.5 Health Aspects

Present knowledge of the constituents in smoke used for food preservation is still far from complete; so no conclusive appraisal of smoking is yet possible. Benzo(a)pyrene (3,4 benzopyrene), which is known to be carcinogenic, is generally regarded as the main substance of toxicological relevance in the smoke used for food preservation and is usually present in the smoke together with other polycyclic compounds. A concentration of 1 microgram per kg of the edible part of the smoked product is regarded as an acceptable maximum quantity. It is possible to keep the concentration of the benzo(a)pyrene below this maximum level if the smoking process is carried out correctly (Kersken 1973, Kersken 1974 b). In cold smoking the quantities of benzopyrene deposited in the food appear to be smaller than in hot smoking. Especially high benzopyrene contents occur in "black" smoking. Combustion temperatures between 300 and 400 °C may be regarded as optimum for the quality of the smoke and the low content of polycyclic hydrocarbons (Kersken 1973). The aldehydes and phenols which also occur in preserving smoke are of minor importance from the toxicological viewpoint.

The smoking of cured meat products can lead to the formation of harmful nitrosophenols (Hofmann 1990). Furthermore after rats were fed simultaneously with liquid smoke condensates, cystein and nitrite, and also after they were fed with smoked foods, such as bacon or cheese, their urine was found to contain *n*-nitrosothiazolidine-4-carboxylic acid as an endogenous nitrosation product (Ikins et al 1988).

25.6 Regulatory Status

In many countries smoking is not subject to any food law controls. In the Federal Republic of Germany the Additives Approval Regulations permit food to be smoked with freshly produced smoke from untreated wood and twigs, heather and cones from coniferous trees. Additionally the use of liquid smoke is approved. Spices may be added or omitted, as desired.

25.7 Antimicrobial Action

The constituents of food-preserving smoke primarily responsible for the antimicrobial action are the aldehydes, low-molecular organic acids and phenol compounds. In addition, the importance of the smoking process in drying out the food should not be underrated. The reduction thus achieved in the water activity of the surface of the smoked food limits the growth of bacteria in particular. Finally, in the case of hot smoking the effect of the heat should be mentioned as a contributory factor.

The antimicrobial action of the smoke constituents increases with a rise in the smoke production temperature (Kersken 1974 a). Compared with other food preservatives, smoke has only a weak antimicrobial effect. In the smoking process only relatively low concentrations of antimicrobially active substances are deposited on the food to be preserved. There are, moreover, a considerable number of microorganisms that are not inhibited by smoke. Thus, unlike cladosporium, mucor and penicillium varieties, for instance, aspergillus strains are largely insensitive (Kersken 1974 a, Asita and Campbell 1990).

25.8 Fields of Use

In southern European countries in particular, smoking is a preserving method used for certain types of hard cheese (in conjunction with other methods).

The main use of smoking is for meat products, especially bacon, ham and sausages of all kinds. In former times, these were smoked in open hearths, but the recent trend in industrial-scale smoking is towards the use of fully air-conditioned maturing and smoking chambers with indirect smoke feed from separate smoke generators. These units enable the temperature, air humidity, smoke feed and fresh air to be accurately controlled.

Smoked herrings, usually known as bloaters or kippers, have been of considerable economic importance, for hundreds of years. Many other varieties of fish are also preserved by smoking, e. g. mackerel, cod, haddock, dogfish, salmon, and eel. The fish may be smoked fresh or else first prepared by salting or other treatment. Depending on the type of fish, cold or hot smoking may be employed (the former for salmon and the latter for bloaters, sprats and mackerel, for instance).

25.9 Other Effects

Smoke not only increases the keeping power of the food it is used to treat but also provides the food with the aroma of the smoke, which is usually desired, and colors the food as well. In many instances smoking is actually employed more for organoleptic reasons than as a preservative treatment, e. g. when used for smoke-flavored potato chips, whisky and smoked beer. Another desirable side effect is the softening action of hot smoking, which is utilized especially in the smoking of fish. Owing to its phenol content, smoke has a slight antioxidative action (Kersken 1974 b), which is due partly to the presence of syringol and 4-methylsyringol (Hrissafidis et al. 1989).

25.10 Literature

Asita AO, Campbell IA (1990) Anti-microbial activity of smoke from different woods. Letters Appl Microbiology 10, 93–95

Hofmann G (1990) Untersuchungen zum Vorkommen von Nitrosophenolen in gepökelten und geräucherten Fleischerzeugnissen. Fleischwirtschaft 70, 1194–1198

Hrissafidis D, Tóth L, Messer S (1989) Gewinnung von wasserfreien Rauchkondensaten für die Überprüfung ihrer antioxidativen Wirkung. Fett Wissenschaft, Technol 91, 207–210

Ikins WG, Gray JI, Mandagere AK, Booren AM, Pearson AM, Stachiw MA, Buckley DJ (1988) Contribution of wood smoke to in vivo formation of n-nitrosothiazolidine-4-carboxylic acid: initial studies. Food Chem Toxicol 26, 15–21

Kersken H (1973) Grundlagen des Räucherns. Räuchern – seit langem bekannt und vielfältig angewandt. Fleischwirtschaft 53, 921, 922, 924 and 925

Kersken H (1974 a) Untersuchungen zur Wirkung von Rauch auf Schimmelpilze. Fleischwirtschaft 54, 1341–1344 and 1346

Kersken H (1974 b) Räuchern heute. Fleischwirtschaft 54, 1721, 1722, 1724 and 1726

Klettner P-G (1979) Heutige Räuchertechnologien bei Fleischerzeugnissen. Fleischwirtschaft 59, 17, 18 and 20–24JA (1987) The flavor chemistry of wood smoke. Food Rev International 3, 139–183

Maga JA (1988) Smoke in food processing. CRC Press, Boca Raton

Münkner W, Meyer C (1994) Untersuchungen zum Einsatz von Flüssigrauch bei der Herstellung geräucherter Fischerzeugnisse. Fleischwirtschaft 74, 547–553

Potthast K (1988) Herstellung und Anwendung von Raucharomen bei Fleischerzeugnissen. Int Z Lebensm-Technol Verfahrenstech 39, 209–213

Potthast K, Eigner G (1986) Formaldehydgehalt in Räucherrauch und in geräuchterten Fleischprodukten. Fleischwirtschaft 66, 1261–1264

Sofos JN, Maga JA, Boyle DL (1988) Effect of ether extracts from condensed wood smokes on the growth of Aeromonas hydrophila and Staphylococcus aureus. J Food Sci 53, 1840–1843

Tóth L (1982) Chemie der Räucherung. Wissenschaftliche Arbeitspapiere der Senatskommission zur Prüfung von Lebensmittelzusatzstoff und -inhaltsstoffen. Verlag Chemie, Weinheim

Tóth L, Potthast K (1984) Chemical aspects of the smoking of meat and meat products. Adv Food Res 29, 87–158

Wittkowski R (1985) Phenole im Räucherrauch. Nachweis und Identifizierung. VCH Verlagsgesellschaft, Weinheim

Chapter 26

Thiabendazole

26.1 Synonyms

IUPAC: 2-(Thiazol-4-yl)-benzimidazole
German: Thiabendazol, 2-(4-Thiazolyl)-benzimidazol. *French:* Thiabendazole. *Italian:* Tiabendazolo. *Spanish:* Tiabendazol. *Russian:* Тиабендаэол.

26.2 History

Thiabendazole has been used in the medical sector for a considerable time as an anthelminthic. In 1964 it was suggested as a fungistatic agent for use in medicine and was later also used to some extent in crop protection. It was from this latter application that thiabendazole came to be adopted for food preservation, where it is used on a limited scale for preserving citrus fruit and bananas.

26.3 Properties

$C_{10}H_7N^3S$, molar mass 201.25, white, crystalline powder that melts at 304–305 °C. Its water solubility reaches a peak of 3.84% at pH 2.2.

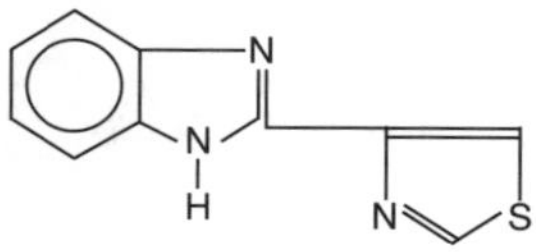

26.4 Analysis

Thiabendazole can be determined in bananas, citrus fruits, and products containing them, by gas chromatography or by HPLC using fluorescence detection (Königer and Wallnöfer 1993, Arenas and Johnson 1994, Oishi et al. 1994). These methods are also suitable for determining thiabendazole in meat products (Wyhowski de Bukanski et al. 1991a and 1991b). Immunological detection methods can also be employed (Bushway et al. 1994).

26.5 Health Aspects

The LD_{50} of thiabendazole for mice, rats and rabbits is between 3.1 and 3.8 g/kg body weight (Robinson et al. 1965). Rats tolerate a daily dosage of 100 – 400 mg/kg body weight in the feed over 30 days. With dosages above 800 mg/kg body weight growth disturbances occur and the mortality rate increases. Additions of 0.1% to the feed have no influence on growth (Robinson et al. 1965). Daily dosages of 200 mg thiabendazole per kg body weight over a period of 180 days do not affect the growth of rats. Even double this quantity only slightly retards their growth, the male rats proving somewhat more sensitive than the females in this regard (Robinson et al. 1965). The same dosage is tolerated by dogs over two years. There has been no evidence of damage to internal organs caused specifically by thiabendazole, apart from minor changes in the blood picture (Robinson et al. 1965).

Thiabendazole has been reported to be teratogenic in mice, however, after administration of extremely large dosages (Ogota et al. 1984). The thiabendazole metabolites, 4- and 5-hydroxythiabendazole and acetylbenzimidazole, may be responsible for these effects (Fujitani et al. 1991). Thiabendazole also has a nephrotoxic effect on mice (Toda et al. 1989 and Toda et al. 1992), particularly where there is a deficiency of glutathione (Mizutoni et al. 1990). In other experiments, however, these data were not confirmed on rats and rabbits (Lankas and Wise 1993).

For the parameters of growth, development, survival period and reproduction, an appropriate NOAEL (no observed adverse effect level) after oral administration to the rat is considered to be 10 mg/kg body weight per day (Wise et al. 1994). The ADI was set at 0–0.3 mg/kg body weight.

26.6 Regulatory Status

Thiabendazole (E 233) is permitted in many countries for the treatment of citrus fruit and bananas. The maximum permissible residual quantities are 6 mg/kg and 3 mg/kg respectively.

26.7 Antimicrobial Action

Thiabendazole is classified among preservatives with a mainly fungistatic action (table 35) (Robinson et al. 1964). It is particularly effective in suppressing molds, especially Penicillium italicum and Penicillium digitatum.

26.8 Fields of Use

The only field of use for thiabendazole in food technology is in the prevention of mold growth on citrus fruit and bananas. Thiabendazole is added in concentrations of 0.1 to 0.45% to wax emulsions or solutions which are used for treating the

Table 35. Inhibitory action of thiabendazole on fungi

Name of test organism	Minimum inhibitory concentration of thiabendazole in ppm
Alternaria solani	< 5
Aspergillus flavus	< 5
Aspergillus fumigatus	< 10
Aspergillus niger	40
Aspergillus glaucus	< 5
Fusarium reticulatum	100
Monilia nigra	10
Mucor mucedo	100
Paecilomyces varioti	100
Penicillium oxalicum	< 5
Penicillium spinulosum	< 5
Rhizopus oryzae	100

fruit (Isshak et al. 1974, Rizk and Isshak 1974). During peeling of citrus fruits treated with thiabendazole, some 5–12% of the residues are transferred to the flesh of the fruit and some 7–14% to the hands. Washing the fruit with hot or cold water is a simple and highly effective way of reducing the transfer of thiabendazole (Königer and Wallnöfer 1990).

26.9 Literature

Arenas RV, Johnson NA (1994) Liquid chromatographic fluorescence method for the determination of thiabendazole residues in green bananas and banana pulp. J AOAC Intern 77, 710–714

Bushway RJ, Young BES, Paradis LR, Perkins LB (1994) Determination of thiabendazole in fruits and vegetables by competitive-inhibition enzyme immunoassay. J AOAC Intern 77, 1243–1248

FAO/WHO (1971) Evaluation of some pesticide residues in food. The Monographs. Issued jointly by FAO and WHO. Rome p. 479–516

Fujitani T, Yoneyama M, Ogata A, Ueta T, Mori K, Ichikawa H (1991) New metabolites of thiabendazole and the metabolism of thiabendazole by mouse embryo in vivo and in vitro. Food Chem Toxicol 29, 265–274

Isshak YM, Rizk SS, Khalil RI, Fahmi BA (1974) Long term storage of Valencia orange treated by thiabendazole. Agric Res Rev 52, 85–98

Königer M, Wallnöfer PR (1990) Übertragung von Thiabendazol-Rückständen beim Schälen von Citrusfrüchten auf Hände und Fruchtfleisch. Dtsch Lebensm Rundsch 86, 251–253

Königer M, Wallnöfer PR (1993) Untersuchungen über das Verhalten von Thiabendazol bei Bananen. Dtsch Lebensm Rundsch 89, 384–385

Lankas GR, Wise DL (1993) Developmental toxicity of orally administered thiabendazole in sprague-dawley rats and New Zealand white rabbits. Food Chem Toxicol 31, 199–207

Mizutoni T, Ito K, Nomura H, Nakanishi K (1990) Nephrotoxicity of thiabendazole in mice depleted of glutathione by treatment with di-buthionine sulphoximine. Food Chem Toxicol 28, 169–177

Ogata A, Ando H, Kubo Y, Hiraga K (1984) Teratogenicity of thiabendazole in ICR mice. Food Chem Toxicol 22, 509–520
Oishi M, Onishi K, Kano I, Nakazawa H, Tanabe S (1994) Capillary gas chromatographic determination of thiabendazole in citrus and apple juices. J AOAC Intern 77, 1293–1296
Rizk SS, Isshak YM (1974) Thiabendazole as a post harvest disinfectant for citrus fruits. Agric Res Rev 52, 39–46
Robinson HJ, Phares HF, Grassle OE (1964) Antimycotic properties of thiabendazole. J Invest Dermatol 42, 479–482
Robinson HJ, Stoerk HC, Grassle OE (1965) Studies on the toxicologic and pharmacologic properties of thiabendazole. Toxicol Appl Pharmacol 7, 53–63
Romminger K, Hoppe H (1976) Beitrag zur Analytik von Thiabendazol auf Zitrusfrüchten und Bananen. Nahrung 20, 407–417
Toda Y, Yoneyama M, Kabashima J, Fujitani T, Nakano M (1989) Effects of thiabendazole on the kidneys of ICR mice. Food Chem Toxicol 27, 307–315
Toda Y, Fujitani T, Yoneyama M (1992) Acute renal toxicity of thiabendazole (TBZ) in ICR mice. Food Chem Toxicol 20, 1021–1030
Tanaka A, Fujimoto Y (1976) Gas chromatographic determination of thiabendazole in fruits as its methyl derivative. J Chromatogr 117, 149–160
Wise LD, Cartwright ME, Seider CL, Sachuk LA, Lankas GR (1994) Dietary two-generation reproduction study of thiabendazole in Sprague-Dawley rats. Food Chem Toxicol 32, 239–246
Wyhowski de Bukanski B, Degroodt J-M, Beernaert H (1991 a) Determination of thiabendazole residues in meat by HPLC using ultraviolet and fluorometric detection. Z Lebensm Unters Forsch 193, 130–133
Wyhowski de Bukanski B, Degroodt J-M, Beernaert H (1991 b) Determination of levamisole and thiabendazole in meat by HPLC and photodiode array detection. Z Lebensm Unters Forsch 193, 545–547

Nisin

27.1 Synonyms

German: Nisin. *French:* Nisine. *Italian:* Nisina. *Spanish:* Nisina. *Russian:* Низин.

27.2 History

As long ago as 1928, it was discovered that a number of *Streptococcus* strains form metabolic products that inhibit other lactic acid bacteria. The possibility of using such pepsin-resistant polypeptides in cheese production was pointed out at the same time (Rogers 1928). After inhibitory substances derived from strains of *Streptococcus* were once again described in 1944 (Mattick and Hirsch 1944), industrial production of nisin commenced in the nineteen-fifties. Nisin has been used in food preservation since that time, although on a limited scale, mainly for improving the keeping properties of processed cheese.

27.3 Commercially Available Forms

Nisin is used exclusively in the form of standardized preparations. Nisaplin, a British product, contains some 2.5% pure substance and is adjusted to 110^6 Reading units/gram.

27.4 Properties

Nisin, $C_{143}H_{230}N_{42}O_{37}S_7$, molar mass 3354.25, is a polypeptide-type antibiotic. It consists of four similarly structured individual components, which in turn are composed of 29 to 34 amino acids, eight of these being sulfur-containing amino acids otherwise found only rarely in nature (Gross and Morell 1971, Hansen 1994). One of these is lanthionin and so substances in this class are also known as lantibiotics. In addition to nisin the group includes subtilin. There are two sub-types of nisin, namely nisin A and nisin Z, which can be distinguished by the amino acid in position 27: aspartic acid in nisin A and histidine in nisin Z.

In its purest form ($40 \cdot 10^6$ Reading units/gram) $4.8 \cdot 10^6$ Reading units dissolve in 1 ml water at pH 2 and $1.6 \cdot 10^6$ Reading units in 1 ml water at pH 5. Commercial preparations form suspensions in water. These are slightly turbid owing to their content of denatured milk protein.

In dry form, nisin remains stable for years. In solution, its stability improves with a decrease in pH value. At pH 2 nisin withstands a temperature of 121 °C for 30 min without losing its activity. At pH values above 4 it decomposes in solutions at a greater or lesser speed, especially if heated. Nisin is particularly sensitive to proteolytic enzymes such as trypsin, pancreatin, salivary enzymes and digestive enzymes but not rennet (de Vuyst and Vandamme 1994).

27.5 Analysis

Determination of the activity of nisin itself and quantitative analysis of nisin in foods are carried out in the plate diffusion test, using *Lactobacillus bulgaricus, L. helveticus* var. *pragensis, Micrococcus flavus* (Vas 1964) or *Bacillus stearothermophilus* (Fowler et al. 1975) as test organisms (Hurst and Hoover 1993).

27.6 Production

Nisin is isolated from cultures of *Streptococcus lactis* (Mattick and Hirsch 1944). Today nisin can also be obtained by genetic engineering (Rauch et al. 1994).

27.7 Health Aspects

27.7.1 Acute Toxicity

Dosages of up to 10^6 Reading units/kg body weight have proved harmless to rats (Frazer et al. 1962).

27.7.2 Subchronic Toxicity

Cheese containing up to $3.33 \cdot 10^6$ Reading units of nisin/kg cheese was fed to rats over 12 weeks without causing damage. The growth and condition of the internal organs corresponded to those of the control animals (Frazer et al. 1962). Doses of 400 mg nisin/kg body weight daily under conditions of stress led to increased mortality in an 8-week feeding experiment with mice (Shtenberg and Ignat'ev 1970).

27.7.3 Chronic Toxicity

In a feeding experiment on rats lasting 24 months and using additions of $3.33 \cdot 10^6$ Reading units/kg feed or corresponding quantities of hydrolyzed, i. e. inactivated, nisin, there were likewise no indications suggesting any kind of harmful effect (Frazer 1962). Similarly, a feeding experiment lasting two years involving the administration of 2 mg nisin/kg body weight daily in combination with 40 mg sorbic acid/kg body weight had no inimical effects on mice or rats (Shtenberg and Ignat'ev 1970). The SCF has set an ADI of 0 – 0.13 mg nisin/kg body weight for a preparation containing 40,000 Reading units (SCF 1990).

27.7.4
Biochemical Behavior

In the digestive tract, nisin is rapidly decomposed enzymatically. Hence, nisin is very unlikely to be toxic to man, as it is present in milk and cheese, and streptococci which produce nisin regularly occur in the intestine (Hawley 1958). The stability of nisin in the intestine is, of course, very poor because of the prevailing pH value and other conditions.

27.8
Regulatory Status

Nisin (E 234) is permitted in some countries chiefly for cheese preservation, including that of processed cheese.

27.9
Antimicrobial Action

27.9.1
General Criteria of Action

The action of nisin is directed against the cytoplasm membrane, which is destroyed immediately after germination of the spores. Consequently the action of nisin against spores is greater than that against vegetative cells (Table 36) (Gould 1964), although nisin cannot be termed directly sporicidal. It appears that, after interaction with the membrane, two nisin molecules form a channel through the membrane, by which free ions can enter or flow out (Winkowski et al. 1995). These effects have a great influence on the proton gradients of the cell membrane. Nisin appears to intensify the heat sensitivity of bacterial spores. There is some reason to believe that the quantity of nisin which acts is that which remains in active form in the mixture after completion of the heating process, this quantity then inhibiting germination of the bacterial spores in the familiar manner (Russell 1971, de Vuyst and Vandamme 1994). Thus, nisin does not attack the spores directly. Its action occurs not during the heating process but afterwards (Tramer 1964).

When combined with chelating agents, such as ethylene diaminotetraacetic acid, or citrates, nisin also inhibits salmonella and *E. coli* (Stevens et al. 1992, ter Steeg 1993).

The action of nisin is recorded internationally in RU (Reading units). IU (International units) are now also commonly used.

Table 36. Inhibitory action of nisin on bacteria (Gould 1964)

Name of test organism	Minimum inhibitory concentration in Reading units nisin/ml on agar after inoculation with	
	spores	vegetative cells
Bacillus aterrimus	2	20
Bacillus coagulans	5	50
Bacillus subtilis	5	100
Bacillus polymyxa	100	100
Bacillus cereus	100	100

1 mg pure nisin is equivalent to 40,000 Reading units or the same number of International units.

The antimicrobial action of nisin is most powerful at pH 6.5 to 6.8, although its stability in this pH range is already very poor. As nisin is destroyed more rapidly by salivary enzymes than the generation time of the bacteria, the flora in the mouth cannot acquire resistance to nisin.

27.9.2
Spectrum of Action

Nisin has a relatively narrow spectrum of action. It acts exclusively against gram-positive bacteria, many lactic acid bacteria, streptococci, bacilli, clostridia and other anaerobic spore-forming microorganisms (Hawley 1957, Ramseier 1960, Vas 1964, Hurst and Hoover 1993). The inhibitory concentrations required for specific bacteria in the nutrient medium test are in the region of 0.25–500 Reading units/gram, i.e. 0.006–12 ppm (Hawley 1957). Yeasts and molds are not inhibited by nisin; indeed, many of these microorganisms tend to decompose it rapidly. *Listeria monocytogenes* can acquire resistance to nisin (Mazzotta and Montville 1995).

27.10
Fields of Use

27.10.1
Dairy Produce

The use of nisin against late blowing of hard cheese is of relatively minor importance compared with its employment for processed cheese, where the primary action of nisin is directed against spore-forming microbes. The main varieties to be suppressed are clostridia, which give rise to white rot, and butyric acid bacteria. The applied concentration is between 2 and 8 mg nisin per kg, i. e. some 100,000 to 300,000 Reading units per kg cheese. Nisin is added during the melting process, as nisin culture or directly in powder form. It can also be added to the raw cheese before melting.

27.10.2
Vegetable Products

Because of its action (even though indirect) in increasing the heat sensitivity of certain bacteria, nisin is used on a small scale as a sterilizing auxiliary. When added to canned products it enables less severe sterilization conditions to be employed, e.g. for tomato products, as well as canned soups, vegetables and mushrooms (Hawley 1957, Vas 1964). In this instance also, its action against clostridia is the predominant factor (Wodsak 1962). Nevertheless the usefulness of adding nisin is disputed.

27.11
General Literature

Lipinska E, Gudkow AW, Karlinkanowa SN (1976) Verwendung von Nisin bei der Käseherstellung. Leipzig: Fachbuchverlag
Hurst A (1978) Nisin: Its preservative effect and function in the growth cycle of the producer organism. Soc Appl Bacteriol Symposium Series No 7, 297–314
Hurst A (1981) Nisin. Adv Appl Microbiol 27, 85–123

27.12
Specialized Literature

Fowler GG, Jarvis B, Tramer J (1975) The assay of nisin in foods. Techn Series Soc Appl Bacteriol 8, 91–105
Frazer AC, Sharratt M, Hickmann JR (1962) The biological effects of food additives. I. – Nisin. J Sci Food Agric 13, 32–42
Gould GW (1964) Effect of food preservatives on the growth of bacteria from spores. Proc 4th Int Symp Food Microbiol Göteborg p. 17–24
Gross E, Morell JL (1971) The structure of nisin. J Amer Chem Soc 93, 4634–4635
Hansen JN (1994) Nisin as a model food preservative. Crit Rev Food Sci Nutrit 34, 69–93
Hawley HB (1957) Nisin in food technology. Food Manuf 32, 270–376 and 430–434
Hawley HB (1958) The permissibility and acceptability of nisin as a food additive. Milchwiss 13, 253–259
Hurst A, Hoover D (1993) Nisin. In: Davidson M, Branen A: Antimicrobials in Food. Dekker, New York, p. 369–394
Mattick ATR, Hirsch A (1944) A powerful inhibitory substance produced by Group N Streptococci. Nature 154, 551
Mazzotta A, Montville T (1995) *Listeria monocytogenes* resistance to Nisin at 10 °C and 30 °C. IFT Ann Meeting '95: Book of Abstracts 81 D-3
Ramseier H (1960) Die Wirkung von Nisin auf *Clostridium butyricum*. Arch Mikrobiol 37, 57–94
Rauch P, Kuipers O, Siezen R, de Vos W (1994) Genetics and protein engineering of nisin. In: Vuyst L, Vandamme E: Bacteriocins of lactic acid bacteria. Microbiology, genetics and applications. Blackie, London, p. 223–249
Rogers LA (1928) The inhibiting effect of *Streptococcus lactis* on *Lactobacillus bulgaricus*. J Bacteriol 16, 321–325
Russell AD (1971) The destruction of bacterial spores. In: Hugo WB: Inhibition and destruction of the microbial cell. Academic Press, London – New York, p. 555–556
SCF (1990): Berichte des Wissenschaftlichen Lebensmittelausschusses, 26th report, EUR 13913 DE

Shtenberg AJ, Ignat'ev AD (1970) Toxicological evaluation of some combinations of food preservatives. Food Cosmet Toxicol 8, 369–380
ter Steeg P (1993) Interactions of nisin, lysozyme and citrate in biopreservation. De Ware(n)-Chemicus 23, 183–190
Stevens KA, Sheldon BW, Klapes NA, Klaenhammer TR (1992) Effect of treatment conditions in nisin inactivation on gram-negative bacteria. J Food Protect 55, 763–766
Tramer J (1964) The inhibitory action of nisin on *Bacillus stearothermophilus*. Proc 4th Int Symp Food Microbiol Göteborg, p. 25–33
Vas K (1964) Anwendung von Nisin in der Lebensmittelindustrie. Dtsch Lebensm Rundsch 60, 63–67
de Vuyst L, Vandamme E (1994) Nisin, a lantibiotic produced by *Lactococcus lactis* subsp. *lactis*: Properties, biosynthesis, fermentation and application. In: de Vuyst L, Vandamme E (ed): Bacteriocins of lactic acid bacteria. Microbiology, genetics and applications. Blackie, London p. 151–221
Winkowski K, Ludescher R, Montville T (1995) Binding affinity of the bacteriocin nisin to lipid bilayers derived from *Listeria monocytogenes* cells. IFT Ann Meeting '95: Book of Abstracts 81 D-4
Wodsak W (1962) Nisin und seine Verwendungsmöglichkeiten. Dtsch Lebensm Rundsch 58, 135–137

Natamycin

28.1
Synonyms

English: Pimaricin, Myprozine, A 5283, Tennecetin. *German:* Natamycin, Pimaricin. *French:* Pimaricine. *Italian:* Pimaricina. *Spanish:* Pimaricina. *Russian:* Пимарицин.

28.2
History

Natamycin was first isolated from culture filtrates of *Streptomyces natalensis* (Struyk and Waisvisz 1955), the antibiotic obtaining the name pimaricin from Pietermaritzburg in Natal Province, South Africa, which was the origin of the soil sample in which the relevant strain had been discovered. The antibiotics A 5283 and tennecetin, which were obtained in the USA from *Streptomyces gilvosporus* and *Streptomyces chattanoogensis* respectively and made available commercially as myprozine, proved to be chemically identical to natamycin.

An important field of use for the antibiotic natamycin is in the medical sector (Raab 1972), but since 1960 or thereabouts efforts have also been made to introduce it for use as a food preservative. Natamycin has achieved a certain practical importance in the preservation of cheese coatings based on polymer emulsions and in the surface preservation of cheeses themselves.

28.3
Commercially Available Forms

Besides the forms of presentation for medical use, namely powder, ointment and vaginal suppositories, natamycin is available as aqueous suspensions and instant powders for the food sector under the names of Delvocid, Delvopos, Delvocoat and Natamax, the last-named being a 50% mixture with lactose.

28.4
Properties

Natamycin, $C_{33}H_{47}NO_{13}$, molar mass 665.73, white crystalline needles which melt at 180 °C upon decomposing. Solubility in water and alcohol is very poor at about 0.005%. Solutions and suspensions are decomposed by oxidizing agents, heavy metals, and light but not by the short-term action of heat (Clar et al. 1964).

Natamycin is stable in the pH range between 4 and 7. In the more strongly acid range the molecule splits apart at the oxygen bridge to form mycosamine, aponatamycin and other substances not yet individually known, some of which dimerize (Brik 1976).

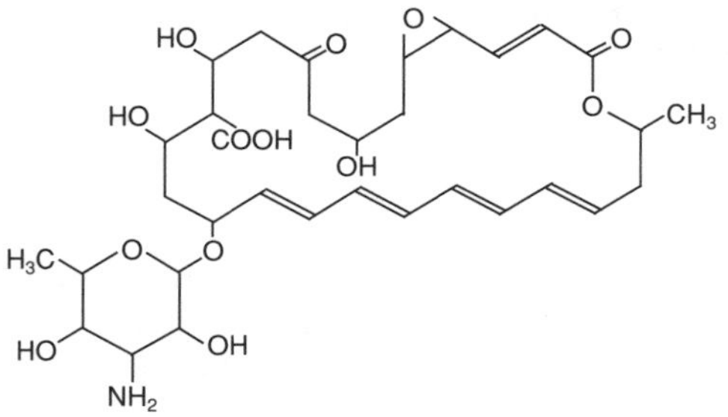

28.5 Analysis

Since natamycin is used in small concentrations and no specific detection tests for natamycin are known, substantial errors are possible in analysis. This encourages its illegal use. As a polyene, natamycin displays absorption maxima at 295, 303 and 311 nm. This fact can be used to assist in its determination (Struyk et al. 1957–1958). There is also a microbiological method based on the agar diffusion test with *Saccharomyces cerevisiae* (Clar et al. 1964). Finally, natamycin can be detected indirectly by using thin-layer chromatography to detect its cleavage products (Millies and Sponholz 1977, Davidson and Doan 1993).

28.6 Production

Natamycin is isolated from cultures of *Streptomyces natalensis*.

28.7 Health Aspects

28.7.1 Acute Toxicity

The LD_{50} of natamycin after peroral administration has been determined as 1.5 g/kg body weight for the mouse (Struyk et al. 1957–1958). For the rat the LD_{50} is 1.5 g/kg body weight (Struyk et al. 1957–1958) but according to other authors (Levinskas et al. 1966) 2.7–4.7 g/kg body weight. For the guinea pig the LD_{50} is 0.45 g/kg body weight (Struyk et al. 1957–1958) and for the rabbit 1.4 g/kg body weight (Levinskas et al. 1966).

28.7.2
Subchronic Toxicity

In dosages of 0.05%, representing 45 mg/kg body weight, natamycin causes no damage in the 90-day feeding test (Levinskas et al. 1966). Higher dosages lead to reduced weight increase, but this reduction is due to vomiting and diarrhea, not organ damage (Struyk et al. 1957–1958, Levinskas et al. 1966).

28.7.3
Chronic Toxicity

Rats tolerate feed containing 0.2% natamycin over a period of two years without appreciable damage. Compared with the control animals, the only difference observed has been a smaller weight increase. In dogs, feed containing 0.05% natamycin likewise causes a reduction in weight increase, but applied concentrations of 0.025% in the feed have no toxic effects (Levinskas et al. 1966).

Since natamycin is an antibiotic whose use in the food sector is basically undesirable, it should be employed only for the surface preservation of cheese and sausage products. A residual quantity of 1 mg/dm^2 with a maximum penetration depth of 5 mm is accepted (SCF 1979). JECFA has set an ADI of 0–0.3 mg/kg body weight (JECFA 1976).

28.8
Regulatory Status

In some countries natamycin (E 235) is permitted for the surface preservation of cheese or to preserve cheese coatings based on plastic. The maximum permitted quantity is 2 mg/dm^2, relative to an outer layer of 5 mm. In some countries natamycin is prohibited for use in foods because it is employed in medicine as an antibiotic.

28.9
Antimicrobial Action

Natamycin has an antimicrobial action on candida similar to that of surface-active substances, i.e. on the cell membrane. It forms a complex with ergosterin, and this increases the permeability of the cell membrane (Oostendorp 1981). The vital activity of the cell can be inhibited by substantially smaller concentrations of natamycin than those required to inhibit the growth of the cells (Bradley et al. 1960).

Natamycin is ineffective against bacteria, viruses and actinomycetes as its action is directed exclusively against yeasts and molds. In particular, it acts against fungi that grow on human skin; hence its use in medicine. Most yeasts and molds are inhibited by natamycin concentrations of 5 to 10 ppm (Klis et al. 1959). Some fungi gradually become resistant to natamycin (Hejzlar and Vymola 1970, Athar and Winner 1971, Dekker and Gielink 1979), although others do not (Boer and Stolk-Horsthuis 1977, Boer et al. 1979).

28.10 Fields of Use

28.10.1 Dairy Produce

Because of its powerful action against molds, natamycin is used in the form of aqueous suspensions for the surface treatment of cheese. As in the case of other preservatives, the foremost consideration in this instance is to restrict the possible formation of mycotoxins by inhibiting mold growth (Kiermeier and Zierer 1975, Bullerman 1977), although natamycin has no direct action against aflatoxin formation or against aflatoxins as such (Kiermeier and Zierer 1975).

The cheese to be protected can be immersed in an approximately 0.05 to 0.25% natamycin suspension, or such a suspension can be used to coat the cheese. The addition of 0.05% natamycin to cheese coatings based on polymer emulsions continues to be an important application (Mol 1966).

Because of its poor water solubility, natamycin penetrates the cheese only slightly, the majority of it remaining concentrated on the cheese surface (Mol 1966, Kiermeier and Zierer 1975). Nevertheless, the extent to which the natamycin migrates into the cheese does depend on the type of cheese in question, a greater depth of penetration having been observed in soft cheese than in hard (Gripon and Bergere 1972). During ripening and storage of the cheese the natamycin is decomposed (Mol 1966, Kiermeier and Zierer 1975). Both the extent that natamycin penetrates the cheese and the speed with which it decomposes are uncertain, as available data are inconsistent. This is due to difficulty in detecting and determining small natamycin concentrations.

28.10.2 Meat Products

The growth of mold on raw sausage can be limited by immersing the sausage in 0.1 to 0.25% natamycin solutions (Hechelmann and Leistner 1969).

28.11 Literature

Athar MA, Winner HI (1971) The development of resistance by *Candida* species to polyene antibiotics in vitro. J Med Microbiol 4, 505–517

Boer E de, Stolk-Horsthuis M (1977) Sensitivity to natamycin (pimaricin) of fungi isolated in cheese warehouses. J Food Protect 40, 533–536

Boer E de, Labots H, Stolk-Horsthuis M, Visser JN (1979) Natamycin-Empfindlichkeit von in Rohwurst herstellenden Betrieben gefundenen Pilzen. Fleischwirtschaft 59, 1887–1889

Bradley SG, Farber PJ, Jones LA (1960) Action of polyene antibiotics on *Candida*. Antimicrob Agents Annu 558–569

Brik H (1976) New high-molecular decomposition products of natamycin (pimaricin) with intact lactone-ring. J Antibiotics 29, 632–637

Bullerman LB (1977) Incidence and control of mycotoxin producing molds in domestic and imported cheeses. Ann Nutrit Aliment 31, 435–446

Clar WK, Shirk RJ, Kline EF (1964) Pimaricin, a new food fungistat. Proc 4th Int Symp Food Microbiol. Göteborg, p. 167–184
Davidson M, Doan C (1993) Natamycin. In: Davidson M, Branen A: Antimicrobials in Foods, Dekker, New York, p. 395–407
Dekker J, Gielink AJ (1979) Acquired resistance to pimaricin in *Cladosporium cucumerinum* and *Fusarium oxysporum* f. sp. *narcissi* associated with decreased virulence. Neth J Plant Pathol. 85, 67–73
Gripon JC, Bergere JL (1972) Etude de la pénétration et de la stabilité de la pimaricine dans les fromages à pâte molle. Lait 52, 428–443
Hechelmann H, Leistner L (1969) Hemmung von unerwünschtem Schimmelpilzwachstum auf Rohwürsten durch Delvocid (Pimaricin). Fleischwirtschaft 49, 639–640
Hejzlar M, Výmola F (1970) Comparative study of pimaricin and fungicidin activity in vitro. J Hyg Epidemiol Microbiol Immunol 14, 211–213
JECFA (1976) Natamycin, WHO Food Additive Series No. 10, Geneva WHO, p. 76–85
Kiermeier F, Zierer E (1975) Zur Wirkung von Pimaricin auf Schimmelpilze und deren Aflatoxinbildung bei Käsen. Z Lebensm Unters Forsch 157, 253–262
Klis JB, Witter LD, Ordal ZJ (1959) The effect of several antifungal antibiotics on the growth of common food spoilage fungi. Food Technol 13, 124–128
Levinskas GJ, Ribelin WE, Shaffer CB (1966) Acute and chronic toxicity of pimaricin. Toxicol Appl Pharmacol 8, 97–109
Millies K, Sponholz WR (1977) Anwendungsmöglichkeiten des Antibiotikums Natamycin (Pimaricin) zur mikrobiologischen Stabilisierung von sauren Getränken. 2. Teil: Eigene Untersuchungsergebnisse. Flüss Obst 44, 56–64
Mol JJ (1966) Ervaringen met een nieuw fungicide bij de schimmelbestrijding op de kaaskorst. Ned Melk Zuiveltijdschr 20, 73–84
Oostendorp JG (1981) Natamycin. Antonie van Leeuwenhoek 47, 170–171
Raab WP (1972) Natamycin (Pimaricin). Its properties and possibilities in medicine. Stuttgart: Georg Thieme
SCF (1979) Bericht des wissenschaftlichen Lebensmittelausschusses, 9th report. Kommission der Europäischen Union, p. 23–24
Struyk AP, Waisvisz JM (Gist-Brocades, Delft) (1955) Pimaricin and process of producing same. US-Patent 3 892 850
Struyk AP, Hoette I, Drost G, Waisvisz JM, Eek T van and Hoogerheide JC (1957–1958) Pimaricin, a new antifungal antibiotic. Antibiot Annu 878–885

Chapter 29

Lysozyme

29.1 Synonyms

German: Lysozym. *French:* Lysozyme. *Italian:* Lisozima. *Spanish:* Lisozima. *Russian:* Лиэоцим.

29.2 History

For decades it has been known that hens' eggs, saliva and other biological materials contain bacteriostatic substances which include an enzyme known as lysozyme (Fleming 1922). Interest in using this bacteriostatic effect for food preservation purposes has existed since the nineteen-seventies in view of the increasing opposition to nitrate. One example of its use is to prevent late blowing in certain hard cheeses.

29.3 Commercially Available Forms

Lysozyme is used in the form of standardized powder preparations.

29.4 Properties

Lysozyme from hens' eggs is an enzyme composed of a peptide chain of 129 amino acid esters which contains four disulfide bonds. Primary and tertiary structures are known. Its molecular weight is around 13 000 – 23 000, depending on origin. Lysozyme is stable in the pH range of around 3 – 7 and is heat resistant up to about 50 °C.

29.5 Analysis

The lysis of *Micrococcus luteus* is used to determine the action of lysozyme and for quantitative analysis of the enzyme in foods (Stelzner et al. 1982).

29.6 Production

Lysozyme is extracted from the white of hens' eggs. It can also be isolated from the culture solutions of certain bacteria and then processed (Hayashi et al. 1981, Conner 1993).

29.7
Health Aspects

Lysozyme is present in many tissues and body fluids. Acute toxicity is not known; rats tolerate 4000 mg/kg body weight (Bianchi 1982). Rabbits tolerate repeated intravenous injections of 500 mg/kg body weight over a period of four weeks more readily than injections of egg albumin (Bianchi 1982). In immunological studies involving guinea pigs, rabbits and humans, lysozyme displayed a lower sensitizing potential than other hen proteins (JECFA 1993). JECFA therefore considers lysozyme acceptable for use in food processing (JECFA 1993).

29.8
Regulatory Status

Lysozyme is approved in various countries as an additive to hard and semi-hard cheese. In the USA lysozyme has GRAS status.

29.9
Antimicrobial Action

Lysozyme is a muramidase. It hydrolyses the glycosidic β_1-U-bond of the murein in the cell wall between the *N*-acetylmuramic acid and the *N*-acetyl glucosamine of the glycopolysaccharides, thereby causing lysis of a number of gram-positive and gram-negative bacteria (Wasserfall et al. 1976). The resistance to the enzyme shown by certain gram-negative bacteria appears to be due to the presence of calcium ions which are necessary for maintaining the stability of the lipopolysaccharide layer. Only when they are removed with EDTA does the murein become susceptible to the dissolving effect of the lysozyme (Wasserfall et al. 1976). The effect of lysozyme is enhanced by hydrogen peroxide, glycine and ascorbic acid (Johansen et al. 1994).

In concentrations of 1 µg/ml, lysozyme does not cause lysis of lactic acid bacteria and coliform organisms. Lysozyme is effective in inhibiting *Clostridium tyrobutyricum*, which often causes late blowing in hard cheese. It is not consistently effective against *Bacillus subtilis* and *Streptococcus lactis*. A favorable effect against *Listeria monocytogenes* can be achieved by reducing the pH (Johansen et al. 1993). Covalent modification of lysozyme with perillaldehyde produces conjugates with a high level of antibacterial activity, for example against *E. coli* and *S. aureus* (Ibrahim et al. 1994).

29.10
Fields of Use

29.10.1
Dairy Produce

Owing to its spectrum of action, lysozyme can be used instead of nitrate to prevent late blowing in certain hard cheeses. Some 500 units/ml is added to the vat milk

(Wasserfall et al. 1976), which is roughly equivalent to 2.5 g crystallized lysozyme/100 l milk. About 90% of the lysozyme forms bonds with the cheese curd and can be regarded as effective. The minimum effective concentration of lysozyme depends very much on the bacteriological quality of the milk used in the cheese dairy.

29.10.2 Meat Products

Tests are documented in which lysozyme was added to sausage meat to improve the storage properties of parboiled sausage (Akashi 1971). However, lysozyme is inactivated in the presence of common salt concentrations exceeding 1.2% (Johansen et al. 1994).

29.10.3 Fish Products

As lysozyme has an effective inhibitory action on clostridia, its use in smoked fish to prevent the reproduction and toxin formation of *Clostridium botulinum* has also been considered.

29.11 Literature

Akashi A (1971) Preservative effect of egg white lysozyme on Vienna sausage. Japan J Zootech Sci 42, 289–295

Bianchi C (1982) Antigenic properties of hen egg white lysozyme (Fleming's lysozyme) and notes on its acute/sub-acute toxicity. Curr Therap Res 31, 494–505

Conner DE (1993) Naturally occurring compounds, Lysozyme. In: Davidson M, Branen A (ed): Antimicrobials in Foods, Marcel Dekker, New York, p. 452–454

Fleming A (1922) On a remarkable bacteriologic element found in tissues and secretions. Proc Royal Soc B 93, 306

Hayashi K, Kasumi T, Kubo N, Tsumura N (1981) Purification and characterization of the lytic enzyme produced by *Streptomyces rutgersensis* H-46. J Agric Biol Chem 45, 2289–2300

Ibrahim M, Hatta H, Fujiki M, Kim M, Yamamoto T (1994) Enhanced antimicrobial action of lysozyme against gram-negative and gram-positive bacteria due to modification with perillaldehyde. J Agric Food Chem 42, 1813–1817

JECFA (1993) WHO Food Additives Series: 30, WHO, Geneva, p. 25–31

Johansen C, Gram L, Meyer AS (1994) The combined inhibitory effect of lysozyme and low pH on growth of *Listeria monocytogenes*. J Food Protect 57, 561–566

Stelner A, Klein U, Klein Y-M, Holtz H (1982) Zur Optimierung des Lysozym-Nachweises: Vergleichende Untersuchungen zur Präparation des Testkeimes *Micrococcus luteus* (*M. lysodeikticus* Fleming). Zbl Bakt Hyg I. Abt Orig A 253, 253–264

Wasserfall F, Voss E, Prokopek D (1976) Versuche über Käsereifung. 5. Die Verwendung von Lysozym anstelle von Nitrat zur Verhinderung der Spätblähung im Käse. Kieler Milchwirtschaftl Forschungsber 28, 3–16

CHAPTER 30

Other Preservatives

Besides the preservatives described in detail in the foregoing chapters, there are some products employed for highly specialized applications. A number of substances have been (or still are) used illegally, while others were employed in the past but are no longer acceptable for toxicological reasons. Others, again, have recently been the subject of detailed discussion and may be of greater importance in future. For the sake of completeness, each group of substances is described briefly in this chapter, which follows the same pattern as the rest of the book.

30.1 Silver

30.1.1 History

According to the Koran, water stored in silver vessels remains fresh for longer (Angehrn 1985). The rediscovery of silver's antimicrobial action dates back to C.F.S. Hahnemann (Bg 1962), whose work appears to have been generally forgotten, since most literature ascribes his findings to the Swiss botanist Nägeli. Hahnemann observed that water would kill microorganisms if it have previously been in contact with native metals, and especially silver. This phenomenon he termed an "oligodynamic" action (Nägeli 1893).

30.1.2 Properties, Analysis

For silver treatment, colloidal silver is used as such or after adsorption by suitable carriers (e.g. titanium dioxide) or in the form of colloidal solutions.

To test for silver, the drink to be investigated or its ash is shaken in a strong sulfuric acid solution together with a solution of dithizone in carbon tetrachloride. The yellow complex substance formed in the presence of silver can be determined photometrically (Miethke and Brösamle 1962).

30.1.3 Health Aspects

No recent systematic investigations have been conducted into the toxicological evaluation of colloidal silver preparations. Earlier literature concludes that the consumption of silver-treated drinks can be regarded as harmless and that, even if silver did accumulate in the body, the risk of argyrosis would be minimal since the silver content of silver-treated drinks is only small.

30.1.4 Regulatory Status

Silver is permitted in some countries for the treatment of drinking water, and in others for drinks and vinegar. The maximum permissible residual quantity after treatment is set at 0.08 mg/l.

30.1.5 Antimicrobial Action

Silver ions intervene in the energy-supplying process of the electron transport chain, where they inhibit cytochromes b and d as well as various flavoproteins (Bragg and Rainnie 1973). There are also descriptions of membrane interactions which disrupt phosphate absorption by the cell (Schreurs and Rosenberg 1982). One possibility being discussed is that the silver ions may be adsorbed on the negatively charged surface of bacteria. Another is that a reaction may occur between silver and enzymes.

The oligodynamic action of silver at pH 8.7 is some three to four times as powerful as at pH 6.3. *Escherichia coli* is killed more rapidly at 25 °C than at 5 °C (Wuhrmann and Zobrist 1958). However, for microorganisms to be killed with some degree of reliability, the action must last for several hours, but this is virtually impracticable and indeed totally so if filtration is employed. Protein, suspended matter, and calcium ions limit the effect of the silver treatment (Wuhrmann and Zobrist 1958, Woodward 1963, Angehrn 1985).

Silver treatment is directed primarily against bacteria. Yeasts and molds are inhibited to a lesser extent (Wuhrmann and Zobrist 1958, Woodward 1963, Antelman 1994).

30.1.6 Fields of Use

Silver treatment is suitable only for liquid products. There are, essentially, three known methods of silver treatment, namely the addition of colloidal silver (Katadyn process, Micropur), filtration using silver-containing filtration media (Sterilit) and the electrolytic deposition of silver in the drink (Elektrokatadyn process, Cumasina).

To disinfect drinking water, the aim is to achieve a silver concentration of 0.025 to 0.1 mg per liter. For other liquids a higher concentration of silver generally has to be used, owing to their possible content of suspended matter. However, since the effect of silver treatment is uncertain even in higher concentrations, silver treatment processes are no longer of any major importance.

30.1.7 Literature

Angehrn M (1985) Einsatz von Silber in der mikrobiologischen Wasseraufbereitung. Brauind 70, 33–37
Antelman M (1994) Silver (II, III) disinfectants. Soap Cosmet Chem Spec 3, 52–59
Bragg P, Rainnie D (1973) The effect of silver ions on the respiratory chain of *Escherichia coli*. Can J Microbiol 20, 833–889
Bg (1962) Hahnemann – Entdecker der oligodynamischen Wirkung von Silber. Dtsch Lebensm Rundsch 58, 73
Miethke H, Brösamle O (1962) Bestimmung von Silber in alkoholfreien Getränken. Dtsch Lebensm Rundsch 58, 71–73
Nägeli C von (1893) Ueber oligodynamische Erscheinungen in lebenden Zellen. Neue Denkschr Allg Schweiz Ges Gesamte Naturwissenschaften 33, 2. Folge, 1–51
Schreurs W, Rosenberg H (1982) Effect of silver ions on transport retention of phosphate by *E. coli*. J Bacteriol 152, 7–13
Woodward RL (1963) Review of the bactericidal effectiveness of silver. J Am Water Works Assoc 55, 881–886
Wuhrmann K, Zobrist F (1958) Untersuchungen über die bakterizide Wirkung von Silber in Wasser. Schweiz Z Hydrol 20, 218–254

30.2 Boric Acid

30.2.1 Commercially Available Forms, Properties, Analysis

Boric acid (H_3BO_3) is employed as such and also in the form of borax, $Na_2B_4O_7 \cdot 10\,H_2O$. Both are moderately water-soluble white lustrous leaflike crystals or white crystalline powders.

In the presence of boric acid, turmeric paper turns orange to red, owing to the formation of rosocyanine. For quantitative determination of boric acid, the food to be investigated is extracted with very dilute hydrochloric acid and titrated with sodium hydroxide solution against phenol phthalein, after which glycerin is added. The glycerin/boric acid complex behaves like a strong acid, thus enabling the boric acid content to be determined by a second titration with sodium hydroxide solution. In a comparative study of various analytical methods the method involving turmeric paper was judged to be the best (Siti-Mizura et al. 1991). In wine, boric acid and its esters can also be detected by ^{11}B-NMR spectroscopy (Lutz 1991).

30.2.2 Health Aspects

The LD_{50} of boric acid is given as 1 to 5 g/kg body weight (Behre 1930, Reith and van Genderen 1956). Other sources quote the LD_{50} for the rat as 5.14 g/kg body weight (Smyth et al. 1969). Acute poisoning is not uncommon with boric acid and borax, especially among children and even when applied externally.

There are no recent results of systematic feeding tests to determine the subchronic toxicity of boric acid and borax. Doses of 0.5 g boric acid administered over a period of 3 to 70 days cause various forms of damage to health (Reith and van Genderen 1956). In reproductive toxicity studies, concentrations of 0.4% boric acid in the feed of mice and rats had a toxic effect on the kidneys and brain of the dams. The incidence of malformed fetuses was significantly increased, as was the mortality (Meindel et al. 1992). Reference is also made to reports of testicular atrophy in experimental animals under the influence of boric acid (SCF 1990). The NOAEL for mutagenic changes in rats is 78 mg/kg feed and for mice 248 mg/kg feed (Price et al. 1990, Meindel et al. 1992).

Borax largely corresponds to boric acid in its biochemical behavior. Both are rapidly and completely absorbed by the body and excreted only slowly (Pfeiffer et al. 1945, Reith and van Genderen 1956). If administered over a lengthy period, e.g. in food, an accumulation of boric acid in the body is likely. In relatively high dosages, boric acid reduces the body's utilization of food and has been used medically as a slimming product on account of this action.

The SCF considers the use of boric acid to preserve caviar to be toxicologically acceptable provided it is employed only within this small field (SCF 1990).

30.2.3 Regulatory Status

Owing to their toxicological properties, boric acid and borax are unimportant today for use as food preservatives. Only for caviar (sturgeon roes) are they still permitted in certain countries up to a maximum quantity of 4 g/kg. In view of the small quantities generally consumed, the risk of boric acid accumulating in the body is not considered to be a problem.

30.2.4 Antimicrobial Action

Boric acid acts by blocking enzymes in the phosphate metabolism. One particular advantage of boric acid is the very low dissociation constant of $7.3 \cdot 10^{-10}$, this being far lower than that of all other preserving acids (von Schelhorn 1952). As a result, boric acid is present almost entirely in an undissociated, i.e. microbiologically effective state, even in the neutral pH range, where preservatives based on carboxylic acids are largely ineffective. For preserving neutral foods, therefore, boric acid is superior to all other preservatives, but because of its relatively low absolute efficacy it has to be used in comparatively high concentrations.

The action of boric acid is directed mainly against yeasts. Its inhibitory action against molds is very slight, while that against bacteria is only partial and leaves some strains completely unaffected (Behre 1952, von Schelhorn 1952).

30.2.5 Fields of Use

Boric acid long retained great importance in Europe as a preservative for margarine and butter. For this application boric acid had the advantage of being readily water-soluble but only slightly fat-soluble and was therefore able to accumulate in the water phase of the emulsified fat, which is the only constituent to be microbiologically susceptible. The applied concentration of 0.5 to 1% was remarkably high in comparison with other preservatives.

Until recently, boric acid was employed in concentrations of around 1% to preserve liquid rennet, whose high pH value makes it difficult to keep fresh with other preservatives. The use of boric acid for liquid egg yolk, meat products, fish products and other sea food is likewise obsolete. Only for caviar is boric acid still used in some instances in concentrations of 0.3 to 0.5%.

30.2.6 Literature

Behre A (1930) Technische Hilfmittel bei der Herstellung von Lebensmitteln. I. Konservierungsmittel. Chem-Ztg 54, 325–327 and 346–347

Behre A (1952) Grundsätzliches zur Konservierungsmittelgesetzgebung. Dtsch Lebensm Rundsch 48, 10–15

Lutz O (1991) Nachweis von Borsäureestern in Wein durch ^{11}B NMR. Naturwiss 78, 67–69

Meindel J, Price C, Field E, Marr M, Myers C, Morrissey R, Schuetz B (1992) Development toxicity of boric acid in mice and rats. Fundam Appl Toxicol 18, 266–277

Pfeiffer CC, Hallman LF, Gersh I (1945) Boric acid ointment, a study of possible intoxication in the treatment of burns. J Amer Med Assoc 128, 266–273

Price C, Field E, Marr M, Myers C (1990) Final report on the developmental toxicity of boric acid in Sprague Dawley rats. Natl Toxicol Program Report NTP 90–155

Reith JF, Genderen H van (1956) De toelaatbaarheid von boorzuur als conserveermiddel in levensmiddelen. Conserva 4, 326–331

SCF (1990) Berichte des Wissenschaftlichen Lebensmittelausschusses. 26th report. EUR 13913 DE

Schelhorn M von (1952) Untersuchungen über Konservierungsmittel. VIII. Wirksamkeit der Borsäure als Konservierungsmittel. Dtsch Lebensm Rundsch 48, 102

Siti-Mizura S, Tee E, Ooi H (1991) Determination of boric acid in foods. Comparative study of three methods. J Sci Food Agric 55, 261–268

Smyth HF, Carpenter CP, Weil CS, Pozzani UC, Striegel JA, Nycum JS (1969) Range-finding toxicity data: List VII. Am Ind Hyg Assoc J 30, 470–476

30.3 Sodium Azide

After derivatization to dinitrobenzoyl chloride, sodium azide can be detected by HPLC (Battaglia and Mitiska 1986).

Sodium azide is a powerful protoplasm poison and therefore not acceptable as a food preservative. Acute oral toxicity for the rat is around 40–60 mg/kg body weight. Sodium azide lowers the blood pressure and has a toxic effect on the central nervous system (Classen et al. 1987). Though described as having no carcinogenic effect, it is said to have mutagenic effects (Dotson and Somers 1989). Neither now nor in the past has this product been permitted in any country for use as a food preservative.

In concentrations of 10 to 20 ppm sodium azide has a powerful inhibitory effect on yeasts. Consequently, on occasions it has been used illegally as a preservative for unfermented fruit juice in concentrations of 5 to 10 g/1000 l and for wine with residual sugar in concentrations of 1 to 3 g/1000 l before more acceptable products with less inimical effects, such as sorbic acid, were approved for this purpose.

30.3.1 Literature

Battaglia R, Mitiska J (1986) Specific detection and determination of azide in wine. Z Lebensm Unters Forsch 182, 501–502

Classen H-G, Elias PS, Hammes WP (1987) Toxikologisch-hygienische Beurteilung von Lebensmittelinhalts- und -zusatzstoffen sowie bedenklicher Verunreinigungen. Berlin Parey, p. 108

Dotson S, Somers D (1989) Differential metabolism of sodium azide in maize callus and germinating embryos. Mutat Res 213, 157–163

30.4 Phosphates

30.4.1 General Aspects

Phosphates are used very successfully in food technology for a number of reasons. Their toxicological properties are well known (Ellinger 1972). Some of them, especially high-polymer phosphates, have a certain antimicrobial action which has been known since as long ago as 1864 (Morgan 1864).

30.4.2 Antimicrobial Action

The action of phosphates is based on that fact that they form a complex linkage with the bivalent metals essential to the microorganism cell, especially magnesium and calcium. By interfering with cell division, this reduces the stability of the cell wall (Post et al. 1963). The effect is most pronounced against bacteria, such as *Staphylococcus aureus, Streptococcus faecalis, Bacillus subtilis* and *Clostridia* (Kelch and Bühlmann 1958, Tompkin 1983, Lee et al. 1994). Phosphates also reduce the heat resistance of a number of bacteria (Hargreaves et al. 1972).

30.4.3 Fields of Use

The antimicrobial effect of the phosphates is of particular importance in the production of processed cheese. The emulsifying salts containing phosphates, which are added in concentrations of 2.5 to 3%, not only convert the raw cheese into an emulsifiable form but also improve its shelf life (Tompkin 1983). AvGard® is the name of an immersing and spraying system based on aqueous trisodium phosphate solutions for reducing salmonella on poultry meat (Kim and Slavik 1994, Lillard 1994).

30.4.4
Literature

Ellinger RH (1972) Phosphates as food ingredients. Cleveland: CRS Press, p. 19–25
Hargreaves LL, Wood JM, Jarvis B (1972) The antimicrobial effect of phosphates with particular reference to food products. The British Food Manufacturing Industries Research Association. Scientific and Technical Surveys. No. 76. Leatherhead: B.F.M.I.R.A.
Kelch F, Bühlmann X (1958) Der Einfluß handelsüblicher Phosphate auf das Wachstum von Mikroorganismen. Fleischwirtschaft 38, 325–328
Kim J-W, Slavik M (1994) Trisodium phosphate (TSP) treatment of beef surfaces to reduce *Escherichia coli* O 157:H7 and *Salmonella typhimurium*. J Food Sci 59, 20–24
Lee R, Hartman R, Stahr M, Olson D, Williams F (1994) Antimicrobial mechanism of long-chained phosphates in *Staphylococcus aureus*. J Food Protect 57, 465–469
Lillard H (1994) Effect of trisodium phosphate on salmonellae attached to chicken skin. J Food Protect 57, 465–469
Morgan J (1864) On a new process of preserving meat. J Soc Arts 12, 347–363
Post FJ, Krishanmurty GB, Flanagan MD (1963) Influence of sodium hexametaphosphate on selected bacteria. Appl Microbiol 11, 430–435
Tompkin RB (1983) Indirect antimicrobial effects in foods: Phosphates. J Food Safety 6, 13–27

30.5
Hydrogen Peroxide

30.5.1
Commercially Available Forms, Properties, Analysis

Hydrogen peroxide, H_2O_2, is marketed chiefly in the form of 3% or 30% aqueous solutions.

Together with freshly prepared aqueous *p*-phenylene diamine solution, hydrogen peroxide causes raw milk to assume a brownish color (peroxidase reaction).

30.5.2
Health Aspects

Hydrogen peroxide, both in pure form and in 30% solution, has a caustic effect. Apart from this, however, it presents no toxicological problems since it rapidly decomposes into water and oxygen in the presence of organic material.

30.5.3
Regulatory Status

Hydrogen peroxide is no longer permitted as a food additive in most countries because it can react as an oxidizing agent with constituents of the food, e.g. vitamins, and also has an undesired bleaching effect. In some countries, hydrogen peroxide may be used only for bleaching starch, gelatin and fish marinades (rollmops).

30.5.4
Antimicrobial Action

Hydrogen peroxide is a disinfectant rather than a preservative since it rapidly kills microorganisms, provided it is employed in the required concentration. Hydrogen peroxide has no long-lasting action, since, once it has acted upon the material to be preserved, it decomposes relatively rapidly within it.

The antimicrobial effect of hydrogen peroxide is based essentially on its oxidative action. This causes all manner of irreversible changes in the microorganism cell. Enzymes, membrane constituents and lipids are unspecifically oxidized and thereby inactivated.

Hydrogen peroxide acts principally against bacteria, whereas yeasts and molds are killed only at relatively high applied concentrations. In nutrient solutions, hydrogen peroxide acts in concentrations as low as 50 to 200 ppm (Kawasaki et al. 1970). *Clostridia* and *Staphylococcus aureus* are more sensitive than aerobic sporeformers and gram-negative bacteria (Amin and Olson 1967, Toledo et al. 1973). The bacteria are killed distinctly faster at 50 °C in the presence of hydrogen peroxide than by heat alone (Amin and Olson 1967). Hydrogen peroxide increases the sensitivity of spores to the effects of heat (Shin et al. 1994).

30.5.5
Fields of Use

Hydrogen peroxide was used to kill the microbes in raw milk in former times when methods of transport and pasteurization were less developed. Known in German as "Buddisierung" (after the inventor, Budde), the method retained a degree of importance for some while. Between 0.02 and 0.05 % hydrogen peroxide is added to the raw milk and left to act for a certain time, after which the excess hydrogen peroxide is destroyed by heating. This process kills pathogenic bacteria and also those causing food spoilage (Budde 1904). By combining hydrogen peroxide with potassium sorbate and using additional cooling, the storage life of milk can be considerably extended (Özdemir and Kurt 1994).

One variant is the treatment of cheese milk with 0.04 to 0.08 % hydrogen peroxide, likewise with the aim of reducing the microbe count in the raw milk. The hydrogen peroxide is allowed to act for 30 min at 50 to 53 °C, after which the excess hydrogen peroxide is destroyed by the action of catalase over a 30 minute period when the milk has been cooled (Roundy 1958). The method is of some importance in the US, where it is known as the peroxide-catalase method.

In tropical countries where unfavorable hygienic conditions exist, treatment with hydrogen peroxide is often the only one by which milk can be preserved from spoilage, at least briefly (Rosell 1957). In these circumstances the drawbacks of this peroxide treatment method (damage to vitamins) are less important than its advantages. Treatment of fresh milk with hydrogen peroxide will be successful only if the milk has a low bacteria count, and should therefore be carried out only after pasteurization (Eapen et al. 1975).

In fish marinades the addition of hydrogen peroxide solution suppresses undesired bacterial changes and odors. The accompanying bleaching effect is not always undesired, especially in the case of herring products.

Hydrogen peroxide is also used for sterilizing packaging materials for drinks, such as beer and milk, in the Tetrapak system, or for gas-phase sterilization of such products (Coles 1995). Excess hydrogen peroxide is destroyed by heating.

30.5.6 Literature

Amin VM, Olson NF (1967) Factors affecting the resistance of *Staphylococcus aureus* to hydrogen peroxide treatments in milk. Appl Microbiol 15, 97–101
Budde CCLG (1904) Ein neues Verfahren zur Sterilisierung der Milch. Tuberculosis 3, 94–98
Coles T (1995) Sterility with peroxide. Manufact Chem March 1995, p. 27–29
Eapen KC, Mattada RR, Sharma TR, Nath H (1975) Keeping quality of fresh milk with hydrogen peroxide as a preservative. J Food Sci Technol 12, 87–90
Kawasaki C, Nagano H, Kono K (1970) Sterilizing effect of hydrogen peroxide in food. Shokuhin Eiseigaku Zasshi 11, 139–142
Özdemir S, Kurt A (1994) Preservation of ewe milk at room and refrigeration temperature by adding hydrogen peroxide and potassium sorbate. Tr J Agric Forestry 18, 219–224
Rosell JM (1957) Die Peroxydkatalase-Behandlung der Milch. Milchwissenschaft 12, 343–348
Roundy ZD (1958) Treatment of milk for cheese with hydrogen peroxide. J Dairy Sci 41, 1460–1465
Shin S, Calvisi E, Beamcin T, Pankratz H, Gerhardt P, Marquis R (1994) Microscopic and thermal characterization of hydrogen peroxide killing and lysin of spores and protection by transition metal ions, chelators and antioxidants. Appl Environm Microbiol 60, 3192–3197
Toledo RT, Escher FE, Ayres JC (1973) Sporicidal properties of hydrogen peroxide against food spoilage organisms. Appl Microbiol 26, 592–597

30.6 Fluorides

Sodium and potassium fluorides inhibit bacteria, yeasts and molds. In former times they were sometimes used without approval for preserving margarine, milk, butter, liquid egg, meat, beer, wine and other foods. Fluorides are also toxic in the concentrations necessary for food preservation, the LD_{50} for the rat being 0.18 g/kg body weight (Smyth et al. 1969). Fluorides inhibit various enzyme systems, besides having a harmful effect on bone and tooth metabolism.

30.6.1 Literature

Smyth HF, Carpenter CP, Weil CS, Pozzani UC, Striegel JA, Nycum JS (1969) Range-finding toxicity data: List VII. Am Ind Hyg Assoc J 30, 470–476

30.7 Bromates

Potassium bromate, $KBrO_3$, used to be employed under the trade name Antibut. It increases the redox potential and thus has an antimicrobial action, especially against bacteria. At one time it was used in concentrations of 0.01 to 0.04% in processed cheese to prevent butyric acid blowing caused by anaerobic spore-forming microorganisms. For toxicological reasons potassium bromate is no longer acceptable (Ballmeier and Epe 1995).

30.7.1 Literature

Ballmeier D, Epe B (1995) Oxidative DNA damage induced by potassium bromate under cell-free conditions and in mammalian cells. Carcinogenesis 16, 335–342

30.8 Ethylene Oxide

30.8.1 Commercially Available Forms, Properties, Analysis

Ethylene oxide, a highly reactive gas with a slightly sweet odor, is usually employed in the form of mixtures with other less explosive and less combustible gases, e.g. carbon dioxide (Cartox, T-Gas, Carboxide, Oxyfume and others) or with fluorinated hydrocarbons (Cry-Oxide).

Interest in ethylene oxide from the analytical viewpoint is focused on the detection and determination of any residues in the foods it is used to treat. For this purpose the fumigated foodstuff is degassed, possibly by heating with xylene. The ethylene oxide driven off can be determined by gas chromatography (Pfeilsticker et al. 1975).

$$
\begin{array}{ccc} CH_2 & — & CH_2 \\ & \diagdown \; O \; \diagup & \end{array}
$$

30.8.2 Health Aspects

The LD_{50} of ethylene oxide for rats and guinea pigs after oral administration is given as some 300 mg/kg body weight (Bruhin et al. 1961, Bruch 1973). Between 100 and 200 mg ethylene oxide per liter respiratory air is fatal to humans in a few instants.

Guinea pigs, rabbits and monkeys tolerate the effect of about 100 ppm ethylene oxide in the respiratory air for 7 hours per day and 5 days a week over a period of several months. Under the same conditions, mice and rats tolerate some 50 ppm without visible harm (Bruch 1961). Other authors state the dose tolerable to laboratory animals over relatively long periods as being about 2 mg per liter air (Phillips and Kaye 1949).

Various authors report a mutagenic action of ethylene oxide (Hogstedt et al. 1979, Anon 1978).

In the case of foods fumigated with ethylene oxide the toxicological risk derives less from the ethylene oxide itself than from its reaction products. The most significant product in quantity terms is ethylene chlorohydrin, the toxicological properties of which have not been fully investigated.

30.8.3
Regulatory Status

Ethylene oxide used to be permitted in some countries for the fumigation of food with a low water content.

30.8.4
Antimicrobial Action

If used in the necessary concentrations, ethylene oxide kills microbes relatively rapidly and is consequently a disinfectant rather than a preservative. Contrary to earlier beliefs, the action of ethylene oxide is certainly not based on the ethylene glycol formed from it by hydrolysis, since the action of the ethylene glycol is much weaker than that of the ethylene oxide itself. Ethylene oxide tends rather to have an alkylating effect on proteins and is thus a general protoplasm poison. In addition, mention should be made of its reactivity with sulfhydryl groups and other active groups of enzyme systems (Phillips 1952, Bruhin et al. 1961).

The effect of the ethylene oxide on microorganisms is governed by the ethylene oxide concentration in the gas space, the duration of action, the temperature at which the action takes place, the pressure and the relative air humidity. By doubling the ethylene oxide concentration, the time required to kill the microbes is halved. The higher the temperature, the faster the action (Driessen and Duin 1975, Toledo 1975). The optimum action of ethylene oxide occurs at 60 °C. Its action can also be accelerated by increasing the pressure (Lammers and Gewalt 1958). Unlike formaldehyde, ethylene oxide is more effective at low relative air humidities than at high values, although the relative air humidity must not fall below a certain minimum value (Bruhin et al. 1961, Hoffmann 1971).

Owing to the general reactivity of the ethylene oxide with the protoplasm protein of microorganisms, the spectrum of action of ethylene oxide is a relatively broad one. Ethylene oxide has a more powerful action against bacteria than against yeasts and fungi, especially Alternaria (Steiger et al. 1974). *Bacillus subtilis*, clostridia and staphylococci are the least sensitive of the bacteria (Driessen and Duin 1975). Ethylene oxide acts rather less powerfully against gram-positive than against gram-negative bacteria.

30.8.5
Fields of Use

Ethylene oxide was originally used to protect stored food from rodents and insects. Since 1933 it has been used to rid foodstuffs of microorganisms (Gross and Dixon 1933).

Owing to its toxicity to humans, its combustibility and its explosive nature, ethylene oxide is used only in enclosed installations. The objects or foods to be treated are brought into contact with ethylene oxide in special sterilization chambers or apparatus. This can be done under vacuum, under normal pressure, or under slightly elevated pressure (Bruhin et al. 1961, Gerhardt 1982). The action is regulated chiefly by controlling its duration and the temperature at which it takes place. The only foods suitable for treatment with ethylene oxide are ones with a low water content, since in these there is little risk that the ethylene oxide will react with the food ingredients (Alguire 1976). The main field of use for ethylene oxide in foods is the fumigation of spices. These can be treated in packaged form with ethylene oxide, since it will penetrate gas-permeable packaging materials. The duration of action is usually a number of hours and the applied concentration some 500 ml ethylene oxide per m^3 air space. A far more important use of ethylene oxide than the destruction of microorganisms in food is the sterilization of medical equipment; but this and the use of ethylene oxide as a pesticide – in cereals, for instance – are largely unrelated to the subject of this book.

30.8.6 Other Effects

Ethylene oxide is capable of reaction with many food constituents to yield products which are undesirable for organoleptic or other reasons. Together with water, ethylene oxide yields ethylene glycol, whilst with alcohols it produces glycol ethers and, with sulfhydril components, thioethers (Hoffmann 1971, Chaigneau 1977, Gerhardt 1982).

30.8.7 Literature

Alguire DE (1976) Regulation of ethylene oxide and propylene oxide in food processing and packaging applications. Food Prod Dev 10: 1, 52–53

Anon (1978) Ethylene oxide, ethylene chlorohydrin and ethylene glycol. Proposed maximum residue limits and maximum levels of exposure. Federal Register 43, 27474–27483

Bruch CW (1961) Gaseous sterilization. Ann Rev Microbiol 15, 245–262

Bruch CW (1973) Sterilization of plastics: Toxicity of ethylene oxide. In: Briggs Philipps G, Miller WS: Industrial Sterilization. Durham: Duke University Press, p. 49–77

Bruhin H, Bühlmann X, Vischer WA, Lammers T (1961) Sterilisation mit Äthylenoxid unter besonderer Berücksichtigung der Anwendung bei Kunststoffen. Schweiz Med Wochenschr 91, 607–613 and 635–639

Chaigneau M (1977) Stérilisation et désinfection par les gaz. Sainte-Ruffine: Maisonneuve, p. 23–107

Driessen FM, Duin H van (1975) Steriliseren met ethyleenoxyde. Voedingsmiddelentechnol 8, 15–19 and 32–33

Gerhardt H, Ladd Effio JC (1982) Äthylenoxidanwendung in der Lebensmittelindustrie. Ein Situationsbericht über "Für und Wider". Fleischwirtschaft 62, 1129–1134

Gross MP, Dixon LF (Liggett & Myers Tobacco Company, New York) (1933) Method of Sterilizing. US Patent 2 075 845

Hoffmann RK (1971) Toxic gases. Ethylene oxide. In: Hugo WB Inhibition and destruction of the microbial cell. London – New York: Academic Press, p. 226–236

Hogstedt C, Malmqvist N, Wadman B (1979) Leukemia in workers exposed to ethylene oxide. J Am Med Assoc 241, 1132–1133

Lammers T, Gewalt R (1958) Ein neues Sterilisationsverfahren mit gespanntem Aethylenoxyd. Z Hyg 144, 350–358

Pfeilsticker K, Fabricius G, Timme G (1975) Simultane, gaschromatographische Bestimmung von Äthylenoxid, Äthylenchlorhydrin und Äthylenglykol in Getreide. Z Lebensm Unters Forsch 158, 21–25
Phillips CR (1952) Relative resistance of bacterial spores and vegetative bacteria to disinfectants. Bacteriol Rev 16, 135–143
Phillips CR, Kaye S (1949) The sterilizing action of gaseous ethylene oxide. Am J Hyg 50, 270–279
Steiger E, Tauchnitz H-D, Löbel A (1974) Über die Resistenz von Pilzen gegenüber Äthylenoxid. Z Gesamte Hyg Ihre Grenzgeb 20, 120–123
Toledo RT (1975) Chemical Sterilants for Aseptic Packaging. Food Technol 29: 5, 102–112

30.9 Glycols

Among the low-molecular glycols, 1,2-propylene glycol has a certain preservative action. Besides having a very low LD_{50}, which is in the region of 20 to 30 g/kg body weight, 1,2-propylene glycol is non-carcinogenic. The product is excreted from the body partly unchanged and partly oxidized to lactic acid, this in turn being utilizable as a source of energy (Informatics 1973).

The antimicrobial action, like that of sodium chloride and sucrose, is based on a reduction in the water activity. Hence, 1,2-propylene glycol has to be used in relatively high concentrations (above 1%). One field of application is in intermediate moisture foods (Davies et al. 1976).

30.9.1 Literature

Davies R, Birch GG, Parker KJ (1976) Intermediate moisture foods. London: Applied Science Publishers, p. 268–269
Informatics, Inc (1973) GRAS (Generally Recognized as Safe) food ingredients – Propylene glycol and derivatives. PB-221 233. Springfield: National Technical Information Service, US Department of Commerce

30.10 Hexamethylenetetramine

30.10.1 History

Hexamethylenetetramine is one of the preservatives deriving originally from the medical sector. Attempts were made to utilize its medically recognized antimicrobial action in the food sector as well, on the assumption that a substance which is effective in the medical sector could not be harmful for foods. As long ago as the beginning of this century, formaldehyde, the actual active substance in hexamethylenetetramine, was used in admixture with hydrogen peroxide to preserve milk. At about that time hexamethylenetetramine began to be used by itself to preserve other foods, mainly of animal origin. In the nineteen-twenties hexamethylenetetramine established itself as a preservative for fish marinades. In most countries it has since been abandoned because of toxicological misgivings.

30.10.2
Properties, Analysis

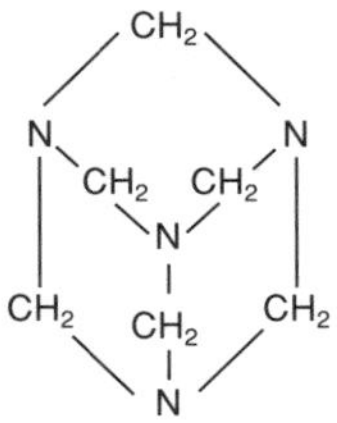

Hexamethylenetetramine is a white crystalline hygroscopic powder with a slightly sweetish flavor, leaving a somewhat bitter aftertaste. In hot water, hexamethylenetetramine is less soluble than in cold water. Aqueous solutions have a slightly basic reaction.

To determine the presence of hexamethylenetetramine, the food to be investigated is subjected to an acid treatment in which formaldehyde is released. Together with chromotropic acid (1,8-dioxynaphthalene-3,6-disulfonic acid), this yields a violet-colored solution, a reaction which can also be employed for quantitative determination of hexamethylenetetramine (Bremanis 1949).

30.10.3
Health Aspects

Hexamethylenetetramine is classed as non-toxic in small doses. Humans can tolerate several grams daily without complications. No LD_{50} is known (Schmidt-Lorenz 1958).

When hexamethylenetetramine was administered in daily dosages of 0.4 g to albino rats in a feeding experiment lasting 90 days, no peculiarities were observed apart from a pronounced and permanent yellow coloration of the fur (Brendel 1964). Doses of 0.06 to 0.125% hexamethylenetetramine to the feed of pregnant beagle bitches over a period of 52 days had no influence on the number, birth weight or health of the litter. In the experiment with the higher dosage, there was a retardation of growth in the first few weeks. Apart from this, no damage was observed within two years; so hexamethylenetetramine is not considered to have any teratogenic action (Hurni and Ohder 1973).

In their action on bacteria (Englesberg 1952) and Drosophila larvae (Rapoport 1946), not only formaldehyde but also hexamethylenetetramine are mutagenic, but there is no agreement as to the significance of this effect on humans (Schmidt-Lorenz 1958, Anon 1964, Natvig et al. 1971).

Addition of 1% hexamethylenetetramine to the drinking water of mice and rats over 60 weeks was likewise tolerated without reaction. Hexamethylenetetramine has no carcinogenic effect (Della Porta et al. 1968). The same results were obtained in a feeding experiment on rats with additions to the diet of 0.16% hexamethylenetetramine (Natvig et al. 1971).

Owing to its good water solubility, hexamethylenetetramine is rapidly absorbed and excreted in the urine, although some is previously converted into formaldehyde by the acid in the gastric juice (Linko and Nikkilä 1959). This, too, is rapidly absorbed by the body. In the blood it immediately attaches to the erythrocytes, which rapidly oxidize it enzymatically to formic acid (Malorny et al. 1965).

The formaldehyde formed from hexamethylenetetramine may react with amino acids and proteins in a variety of ways. Of the formylated proteins produced, some can be utilized by the body effectively and others less so (Lang 1951, Schmidt-Lorenz 1958).

30.10.4
Regulatory Status

Hexamethylenetetramine used to be employed in some countries for preserving fish products, especially in northern and central Europe.

Its use as a food preservative has now been largely discontinued because of misgivings concerning the toxicity of the formaldehyde that evolves from it. In some countries it is still permitted for one application, namely the preservation of provolone cheese.

30.10.5
Antimicrobial Action

Hexamethylenetetramine has no antimicrobial action of its own. Its microbicidal effect is based on that of the formaldehyde released from hexamethylenetetramine in an acid medium. The higher the acidity of the medium, the more powerful is the hydrolytic cleavage and the greater the antimicrobial action.

The action of the formaldehyde and thus of the hexamethylenetetramine is based on the reaction with proteins in the microorganism cell. The same chemical reaction also leads to inactivation of enzymes such as dehydrogenases (Linko and Nikkilä 1959). Proteins and foods or food constituents containing proteins may impair the antimicrobial effect of hexamethylenetetramine. Consequently, hexamethylenetetramine has to be used in higher concentrations in commercial-scale food preservation than would be expected on the basis of nutrient medium tests.

In view of the unspecific action of formaldehyde against protein, the antimicrobial effect of hexamethylenetetramine is fairly universal if the pH conditions are favorable (Nikkilä and Linko 1958). Bacteria are, however, inhibited by slightly lower concentrations than are yeasts (von Schelhorn 1954). The least powerful effect of hexamethylenetetramine is that against molds (von Schelhorn 1954).

30.10.6
Fields of Use

An addition of hexamethylenetetramine to cheese milk prevents certain forms of late blowing in hard cheese caused by bacteria. Only in the case of provolone and other Italian hard cheeses is the process of any importance, however. The formaldehyde content of the cheese is some 20 to 25 mg/kg.

Because the formaldehyde evolved from hexamethylenetetramine is highly effective against lactobacilli and other spoilage microorganisms or pathogenic bacteria, it was used for a long time to preserve fish marinades, anchovies, caviar, mussel and shrimp products, and processed crab. It was employed mainly in combination with benzoic acid and/or sorbic acid in order to suppress the growth of yeasts and molds against which hexamethylenetetramine is ineffective on its own. The applied concentration in fish marinades was between 0.02 and 0.03%, depending on the pH value.

Because of misgivings concerning the toxicity of the formaldehyde evolved from hexamethylenetetramine, today it is of virtually no importance for preserving fish. Technology has also rendered it superfluous: it has been replaced by heating and cooling processes.

30.10.7
Literature

Anon (1964): Give a dog a bad name ... Food Cosmet Toxicol 2, 745–749

Bremanis E (1949) Die photometrische Bestimmung des Formaldehyds mit Chromotropsäure. Z Anal Chem 130, 44–47

Brendel R (1964) Untersuchungen an Ratten zur Verträglichkeit von Hexamethylentetramin. Arzneim Forsch 14, 51–53

Della Porta G, Colnaghi MI, Parmiani G (1968) Non-carcinogenicity of hexamethylenetetramine in mice and rats. Food Cosmet Toxicol 6, 707–715

Englesberg E (1952) The mutagenic action of formaldehyde on bacteria. J Bacteriol 63, 1–11

Hurni H, Ohder H (1973) Reproduction study with formaldehyde and hexamethylenetetramine in beagle dogs. Food Cosmet Toxicol 11, 459–462

Lang K, Frimmer M, Bernert D (1951) Stoffwechselverhalten und Verträglichkeit formylierter und acetylierter Proteine. Z Gesamte Exp Med 117, 288–296

Linko RR, Nikkilä OE (1959) Chemical preservatives in foodstuffs. III. Hexamethylenetetramine as mold inhibitor and the antagonistic action of amino acids. Maataloustieteellinen Aikakauskirja 31, 162–173

Malorny G, Rietbrock N, Schneider M (1965) Die Oxydation des Formaldehyds zu Ameisensäure im Blut, ein Beitrag zum Stoffwechsel des Formaldehyds. Naunyn-Schmiedebergs Arch Exp Pathol Pharmakol 250, 419–436

Natvig H, Andersen J, Wulff Rasmussen E (1971) A contribution of the toxicological evaluation of hexamethylenetetramine. Food Cosmet Toxicol 9, 491–500

Nikkilä OE, Linko RR (1958) Chemical preservatives in foodstuffs. II. The effect on moulds. Maataloustieteellinen Aikakauskirja 30, 125–131

Rapoport IA (1946) Carbonyl compounds and chemical mechanisms of mutations. Dokl Akad Nauk SSSR 54, 65–67

Schelhorn M von (1954) Untersuchungen über Konservierungsmittel. IX. Hexamethylenetetramin als Konservierungsmittel. Dtsch Lebensm Rundsch 50, 90–92

Schmidt-Lorenz W (1958) Zur Verwendbarkeit von Hexamethylenetetramin und Formaldehyd als Konservierungsmittel. Z Lebensm Unters Forsch 108, 423–441

30.11
Monochloroacetic Acid

In a long-term experiment with 0.1% monochloroacetic acid in the feed of rats, growth retardation occurred but there were no indications of any carcinogenic action (Fuhrmann et al. 1955). Monochloroacetic acid is not permitted for food preservation in any country. Its action is only weak and directed more against yeasts than bacteria and molds. Monochloroacetic acid and its sodium salt have occasionally been used in some countries on a small scale for the stabilization of wine.

30.11.1 Literature

Fuhrmann FA, Field J, Wilson RH, DeEds F (1955) Monochloracetate: Effects of chronic administration to rats on growth, activity, and tissue metabolism and inhibitory effects in vitro compared with monoiodacetate and monobromacetate. Arch Intern Pharmacodyn 102, 113–125

30.12 Monobromoacetic Acid

From the nineteen-forties onwards monobromoacetic acid and its esters with ethanol, benzyl alcohol and glycols were marketed under various names for the preservation of drinks, especially beer, wine and juices, although in no country have they at any time been granted approval. For a long time they were a classic example of illegal preservatives, a status they acquired because of their strong antimicrobial action combined with instability, which makes them difficult to detect. In the mid 1980s mono*bromo*acetic acid was found in Bavaria to have been added illegally to beer. In wine, a quantity as small as 30–50 mg/l is highly effective against secondary fermentation by yeast.

The LD_{50} of monobromoacetic acid for mice is 100 mg, and for rats 50 mg/kg body weight (Fuhrmann et al. 1955). In pigs, additions of 10 to 54 mg/kg body weight over a period of 28 to 105 days proved highly toxic, whereas the administration of 2 to 6 mg/kg body weight over 350 to 450 days caused no accumulation, injury to the health or interference with growth and reproduction (Dalgaard-Mikkelsen et al. 1955). Monobromoacetic acid is not carcinogenic (Dalgaard-Mikkelsen et al. 1955).

The antimicrobial action of monobromoacetic acid and its derivatives is based on the reaction with SH enzymes, although it is probable that oxidation and reduction processes in the cell are also affected. This action is directed chiefly against yeasts. Bacteria and molds are inhibited only by relatively large dosages.

30.12.1 Literature

Dalgaard-Mikkelsen S, Kvorning SA, Møller KO (1955) Toxic effects of monobromoacetic acid on pigs. Acta Pharmacol Toxicol 11, 13–32

Fuhrmann FA, Field J, Wilson RH, DeEds F (1955) Monochloracetate: Effects of chronic administration to rats on growth, activity and tissue metabolism and inhibitory effects in vitro compared with monoiodacetate and monobromacetate. Arch Intern Pharmacodyn 102, 113–125

30.13 Lactic Acid

30.13.1 General Aspects

Lactic acid is a colorless to slightly yellowish liquid with unlimited miscibility with water. The LD_{50} of lactic acid after peroral administration is 3.7 g/kg body weight for the rat and 1.8 g/kg body weight for the guinea pig (Spector 1956). Lactic acid is not mutagenic (Life Science Research Office 1978). As a naturally occurring product which forms in foods during fermentation processes, lactic acid is subject to only a few food law restrictions.

30.13.2 Antimicrobial Action

The antimicrobial action of lactic acid is relatively slight, a preservative effect occurring only at concentrations above 0.5%. The action is directed chiefly against anaerobic bacteria. One important aspect of this effect is the reduction it brings about in the pH. As many yeasts and molds are capable of utilizing lactic acid in their metabolism, it is frequently combined with other preservatives such as benzoic and/or sorbic acid (Woolford 1975) as well as natamycin in order to protect foods from spoilage by these particular microorganisms.

30.13.3 Fields of Use

Lactic acid is among the oldest known preservatives, its effects having been utilized for centuries in the production of sauerkraut, pickles, beans and olives. Especially in the Far East and eastern Asia, other foodstuffs, including animal products, are also preserved by lactic acid fermentation. The souring of milk and cream, as well as some aspects of cheese manufacture, are further instances of lactic acid's use as a preservative. The lactic acid produced in these foods is formed by fermentation from the carbohydrates which the foods contain; but, since completely new foods are formed by the wide variety of biochemical reactions that occur (much as in the production of alcoholic drinks by fermentation of juices) the process is not, strictly speaking, a chemical method of food preservation. Recently lactic acid and lactates have been recommended for preserving meat and sausage products (Shelef 1993, Weaver and Shelef 1994).

30.13.4 Literature

Life Science Research Office (1978) Evaluation of the health aspects of lactic acid and calcium lactate as food ingredients. PB-283 713. Springfield National Technical Information Service. US Department of Commerce

Shelef L (1994) Antimicrobial effects of lactates: A review. J Food Protect 55, 445–450

Spector WS (1956) Handbook of toxicology. Volume 1. WB Saunders, Philadelphia – London, p. 262–263
Weaver A, Shelef L (1993) Antisterial activity of sodium, potassium and calcium lactate in pork liver sausages. J Food Safety 13, 133–146
Woolford MK (1975) Microbiological screening of food preservatives, cold sterilants and specific antimicrobial agents as potential silage additives 26, 229–237

30.14 Glycerin Esters of Medium-chain Fatty Acids

In the nineteen-thirties many studies were undertaken into the antimicrobial action of medium-chain fatty acids (Tetsumoto 1933). It may have been these studies that led to the discovery of the antimicrobial action of sorbic acid in 1939. It was later discovered that the glycerin esters of medium-chain fatty acids, especially monoglycerides, have not only a powerful surface activity but also a much more powerful antimicrobial action than the fatty acids themselves. The glycerin esters of laurinic acid, known as lauricidin or monolaurin, have the most powerful antimicrobial activity (Kabara 1993).

The esters have a very broad spectrum of action against bacteria, yeasts and molds. Mixtures of sorbic acid with various esters of medium-chain fatty acids have been used experimentally in the US as food preservatives (Razavi-Rohani and Griffiths 1994). However, the esters seem to have organoleptic and other technological drawbacks, since they have so far not attained any great practical importance. In the US lauricidin has GRAS status, but only as an emulsifier for foods. It is not permitted for use as a preservative, either there or in the European Union. The main field of use is in the cosmetics sector, which utilizes both its emulsifying and preserving properties.

30.14.1 Literature

Kabara JJ (1984) Lauricidin. The nonionic emulsifier with antimicrobial properties. In: Kabara JJ: Cosmetic and drug preservation. Principles and practice. Marcel Dekker, New York, p. 305–322
Kabara JJ (1993) Medium-chain fatty acids and esters. In: Davidson PM, Branen AL: Antimicrobials in Foods. Marcel Dekker, New York, p. 307–342
Razavi-Rohani SM, Griffiths MW (1994) The effect of mono- and polyglycerol laurate on spoilage and pathogenic bacteria associated with foods. J Food Safety 14, 131–151
Tetsumoto S (1933) Sterilizing action of acids. II. Sterilizing action of saturated fatty acids. J Agric Chem Soc Japan 9, 388

30.15 Ethylene Diamine Tetraacetic Acid

Known for short as EDTA, ethylene diamine tetraacetic acid is used as a synergist for antioxidants and preservatives. The forms of EDTA used are its sodium and calcium salts, on account of their complex-forming effect. In feeding experiments EDTA proved to have little toxic effect even in excess concentrations (Krum and Fellers 1952).

EDTA is, however, capable of forming complexes with vital bivalent and multivalent metals in the body, thereby withdrawing them from the metabolism.

In concentrations exceeding around 300 ppm, ethylene diamine tetraacetic acid inhibits the growth of bacterial spores (Bulgarelli and Shelef 1985), its action being purely antibacterial and scarcely affecting either yeasts or molds at all (Russel 1971). By forming complexes with bivalent metal ions EDTA increases the permeability of the cell membranes to other preservatives and thus has a synergistic effect.

In some countries ethylene diamine tetraacetic acid is used on a small scale for the preservation of shrimps. Dips of the sodium salt of EDTA suppress the formation of trimethylamine on chilled fish fillets (Levin 1967).

30.15.1 Literature

Bulgarelli MA, Shelef LA (1985) Effect of ethylenediamine-tetraacetic acid (EDTA) on growth from spores of *Bacillus cereus*. J Food Sci 50, 661–664

Krum JK, Fellers CR (1952) Clarification of wine by a sequestering agent. Food Technol 6, 103–106

Levin RE (1967) The effectiveness of EDTA as a fish preservative. J Milk Food Technol 30, 277–283

Russel AD (1971) Ethylenediaminetetra-acetic acid. In: Hugo WB (ed) Inhibition and destruction of the microbial cell. Academic Press, London p. 209–224

30.16 Allyl Mustard Oil

Allyl isothiocyanate, $CH_2=CH-CH_2-N=C-S$, molar mass 99.16, is an oil with poor water solubility and a disagreeable odor. It is a natural constituent of mustard, horseradish and other plants.

If administered in relatively high concentrations it causes damage to the bladder. It is mutagenic (Neudecker and Henschler 1985) and carcinogenic for the rat (NTP 1982).

Allyl mustard oil is approved in Italy in the form of paraffin wax tablets for the stabilization of wine in relatively large containers.

30.16.1 Literature

Neudecker T, Henschler D (1985) Allyl isothiocyanate is mutagenic in *Salmonella typhimurium*. Mutation Res 156, 33–37

NTP (1982) Technical report on the carcinogenesis bioassay of allyl isothiocyanate (CAS No. 57–06–7) in F 344/N rats and $B6C_3F_1$ mice (Gavage Study) NTP-81–38. NIH Publication No. 83–1790. US Department of Health and Human Services

30.17 Thiourea

Thiourea is effective primarily against molds and to a lesser extent against bacteria. For a while it was employed in the form of 2 to 10% aqueous solutions or wax emulsions containing 4 to 6% thiourea to preserve citrus fruit, but for toxicological reasons (carcinogenicity) it is now obsolete as a food preservative.

30.18 Dehydroacetic Acid

30.18.1 Properties, Analysis

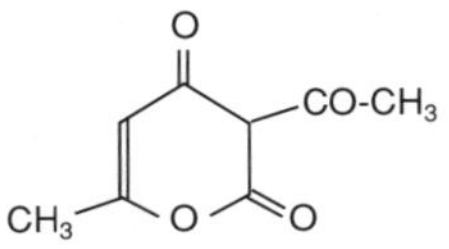

Dehydroacetic acid is sparingly water-soluble. Its sodium salt has better water solubility.

Dehydroacetic acid can be isolated from the food under investigation by shaking with organic solvents such as ether or petroleum ether. For qualitative detection, use can be made of the deep blue coloration of the copper salt. Like sorbic acid and benzoic acid, dehydroacetic acid can be detected by thin-layer chromatography (Khan et al. 1994). Quantitative determination methods can employ the orange-red coloration assumed by dehydroacetic acid in an alkaline medium with salicylaldehyde or the absorption maximum at 312 nm (Woods et al. 1950).

30.18.2 Health Aspects

The LD_{50} of dehydroacetic acid for rats after administration per os in the feed in the form of an oily suspension is 1000 mg/kg body weight (Spencer et al. 1950). The corresponding value for the sodium salt of dehydroacetic acid is 570 mg/kg for rats and 400 mg/kg for dogs (Seevers et al. 1950).

Rats tolerate 0.1 g dehydroacetic acid per kg body weight over a period of 34 days without complications, whereas repeated doses of 0.3 g/kg body weight lead to relatively severe weight losses and damage to various internal organs (Spencer et al. 1950). Over a period of 105 days a feed containing 0.2% dehydroacetic acid leads to a weight increase less than that of control animals on a normal diet (Ritschel 1965).

A two-year experiment in which rats were fed a diet containing 0.1% dehydroacetic acid gave no indication of damage due specifically to the preservative (Spencer et al. 1950). Monkeys tolerate 0.1 g dehydroacetic acid per kg body weight five times per week over 1 year without damage, although 0.2 g/kg body weight leads to growth disturbances and pathological changes to organs (Spencer et al. 1950). The dosage tolerated by dogs for long periods is 50 mg/kg body weight (Seevers et al. 1950). Humans tolerate 9 mg dehydroacetic acid per kg body weight for 173 days, the plasma concentrations that occur during the experiment being 10–15 mg/100 ml (Shideman et al. 1950).

Dehydroacetic acid is rapidly and completely absorbed by the human and animal organisms. The substance is dissipated in the plasma and numerous organs (Woods et al. 1950) and inhibits various oxidation enzymes (Seevers et al. 1950). It is excreted in the urine, albeit rather slowly (Shideman et al. 1950) and so regular consumption is likely to result in accumulation. Owing to the reactivity of the keto group in the acetyl side chain of the molecule and the possibility that dehydroacetic acid will link with amino acids, it is not considered suitable as a preservative for foods (Barman et al. 1963).

30.18.3
Regulatory Status

Dehydroacetic acid and its sodium salt are permitted for use only in a few east Asian countries for preserving certain foods. In Europe dehydroacetic acid is not permitted as a food preservative.

30.18.4
Antimicrobial Action

Dehydroacetic acid has a relatively low dissociation constant and therefore remains effective even in the high pH range.

The action of dehydroacetic acid is directed primarily against yeasts and molds, the minimum inhibitory concentrations being around 50 to 500 ppm (Wolf 1950, von Schelhorn 1952). Far higher dosages are needed to suppress the growth of bacteria. Minimum inhibitory concentrations are between 1000 and 4000 ppm (Brodersen and Kjaer 1946, Wolf 1950).

30.18.5
Fields of Use

Treatments with aqueous solutions of sodium dehydroacetate can protect peeled or cut squash from undesired mold attack. Dehydroacetic acid has also been recommended for foods with a high pH value which are difficult to preserve, e.g. baked goods, cheese and margarine, as well as for producing fungistatic packaging materials. Because of its toxicity, however, dehydroacetic acid has been unable to achieve major importance in any of these fields of use.

30.18.6
Literature

Barman TE, Parke DV, Williams RT (1963) The metabolisms of dehydroacetic acid (DHA). Toxicol Appl Pharmacol 5, 545–568

Brodersen R, Kjaer A (1946) The antibacterial action and toxicity of some unsaturated lactones. Acta Pharmacol 2, 109–120

Khan S, Murawski M, Sherman J (1994) Quantitative HPTLC determination of organic acid preservatives in beverages. J Liq Chrom 17, 855–865

Ritschel WA (1965) Zur Verträglichkeit der Dehydracetsäure. Arzneim Forsch 15, 220–222

Schelhorn M von (1952) Die Dehydracetsäure als Konservierungsmittel für Lebensmittel. Dtsch Lebensm Rundsch 48, 16–18
Seevers MH, Shideman FE, Woods LA, Weeks JR, Kruse WT (1950) Dehydroacetic Acid (DHA). II. General pharmacology and mechanism of action. J Pharmacol Exp Ther 99, 69–83
Shideman FE, Woods LA, Seevers MH (1950) Dehydroacetic acid (DHA). IV. Detoxication and effects on renal function. J Pharmacol Exp Ther 99, 98–111
Spencer HC, Rowe VK, McCollister DD (1950) Dehydroacetic acid (DHA). I. Acute and chronic toxicity. J Pharmacol Exp Ther 99, 57–68
Wolf PA (1950) Dehydroacetic acid a new microbiological inhibitor. Food Technol 4, 294–297
Woods LA, Shideman FE, Seevers MH, Weeks JR, Kruse WT (1950) Dehydroacetic Acid (DHA). III. Estimation, absorption and distribution. J Pharmacol Exp Ther 99, 84–97

30.19 Salicylic Acid

30.19.1 Properties, Analysis

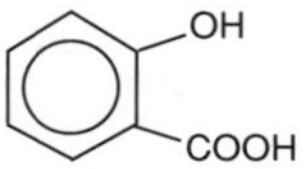

Salicylic acid is a white, sparingly water-soluble crystalline powder.

With iron-(3)-chloride, salicylic acid produces a violet coloration, which can be used in quantitative determination.

30.19.2 Health Aspects

The acute oral toxicity of salicylic acid for rabbits is given as 1.1 to 1.6 g/kg body weight. In dogs the minimum lethal dose is 0.45–0.5 g/kg body weight (Spector 1956).

Salicylic acid in a dosage of 0.1 g/kg body weight, administered daily to mice over a period of 2 months, causes no damage. Organ damage is, however, produced by 0.3 g/kg body weight (Herz and Stampfl 1951). Salicylic acid, administered in the feed in the form of its sodium salt, has no teratogenic action (Minor and Becker 1971).

The discoverer of salicylic acid's antimicrobial action reported an experiment on himself, in which he took approximately 1 g salicylic acid daily for two years without harm (Kolbe 1878). No systematic investigations are known to have been conducted recently into the chronic toxicity of salicylic acid.

Salicylic acid is absorbed rapidly and completely by the body and excreted as a reaction product with glycocoll (salicyluric acid) or in another compound form (Herz and Stampfl 1951). Since excretion is only slow, a risk of accumulation exists. Salicylic acid can have a sensitizing action after oral ingestion.

30.19.3 Regulatory Status

Salicylic acid is of virtually no importance today as a food preservative.

30.19.4
Antimicrobial Action

By reacting with protein, salicylic acid damages the plasma of the microorganism cells. It most likely also intervenes in enzyme reactions but there have been no recent systematic investigations into the preservative effect of salicylic acid. However, one factor of particular importance to the antimicrobial action of salicylic acid is its influence on the formation of pantothenic acid, which is essential to life for many microorganisms, especially bacteria (Wyss 1948). Owing to its dissociation behavior, salicylic acid can be used only for highly acid foods, its action in this regard being even less favorable than that of formic acid and benzoic acid.

Salicylic acid is more effective against fungi and yeasts than against bacteria. The antibacterial action of salicylic acid is, however, better than that of benzoic acid because of the former's phenolic OH group (von Schelhorn 1951).

30.19.5
Fields of Use

Until the middle of the twentieth century, salicylic acid was one of the most commonly used preservatives for foods. It was employed to preserve liquid egg yolk, fish marinades and gherkins pickled in either brine or vinegar, as well as olives and fruit products. The last major field of use in Germany was the preservation of fruit, compotes, marmalades and jams in the household, the sparingly water-soluble salicylic acid being added to the heated preserve. Marmalades, jams and jellies were sometimes additionally protected by a sheet of parchment previously immersed in alcoholic solutions of salicylic acid and placed on their surface.

Originally introduced as an alternative to benzoic acid, salicylic acid was superseded, first by benzoic acid and later by sorbic acid, because of its side effects and toxicological characteristics. It is no longer of any importance in the food sector.

30.19.6
Other Effects

As salicylic acid readily splits off phenol at elevated temperatures, changes in odor and flavor may occur in foods subjected to heat treatment in the course of their preparation or processing.

Foods containing salicylic acid may be discolored if metals are present, especially iron.

30.19.7
Literature

Herz A, Stampfl B (1951) Verträglichkeit oft wiederholter kleiner Salicylsäuregaben. Z Gesamte Exp Med 118, 76–90

Kolbe H (1878) Ist anhaltender Genuß kleiner Mengen Salicylsäure der Gesundheit nachteilig? Z gegen Verfälsch Lebensm u sonst Verbrauchsgegenstände 1, 370–371

Minor JL, Becker BA (1971) A comparison of the teratogenic properties of sodium salicylate, sodium benzoate and phenol. Toxicol Appl Pharmacol 19, 373
Schelhorn M von (1951) Untersuchungen über Konservierungsmittel. V. Zur Frage des Vergleichs der Wirksamkeit von Konservierungsmitteln. Dtsch Lebens Rundsch 47, 16–18
Spector WS (1956) Handbook of toxicology. Volume 1. WB Saunders, Philadelphia – London, p. 262–263
Wyss O (1948) Microbial inhibition by food preservatives. Adv Food Res 1, 373–393

30.20 p-Chlorobenzoic Acid

p-Chlorobenzoic acid can be identified and quantitatively determined by an HPLC method (Flak and Schaber 1988).

Acute and subchronic toxicity of *p*-chlorobenzoic acid are somewhat poorer than those of benzoic acid. Rats tolerate feed containing 0.2% *p*-chlorobenzoic acid for five months without untoward effects on growth, food utilization, reproductive capacity, state of health and histology of the liver and kidneys (Kieckebusch et al. 1960). By and large, *p*-chlorobenzoic acid corresponds to benzoic acid in its behavior. It is largely eliminated from the body in the form of *p*-chlorohippuric acid.

In its antimicrobial action too, *p*-chlorobenzoic acid is similar to benzoic acid. This action is related to the undissociated molecule and is therefore heavily dependent on the pH value of the substance to be preserved. In this respect *p*-chlorobenzoic acid behaves somewhat less favorably than benzoic acid and can therefore be used only to preserve foods with a moderately acid reaction. Its action against molds tends to be poorer than that of benzoic acid.

For reasons of solubility, *p*-chlorobenzoic acid was used chiefly in the form of its sodium salt. Its main fields of application were originally fruit products such as juices, fruit purees and pulps. Initially it was assumed that the product had the advantage over formic acid, sulfur dioxide and benzoic acid of separating out completely in the highly acid pH range and then forming a precipitate which remained behind when the food was eaten (Strahlmann 1974). Later the sodium salt of *p*-chlorobenzoic acid was used increasingly as a preservative for fish products. The applied concentration was around 0.05 to 0.1%. *p*-Chlorobenzoic has ceased to be employed as a preservative today and is no longer approved.

30.20.1 Literature

Flak W, Schaber R (1988) Die Bestimmung von Konservierungsmitteln in Wein und anderen Getränken mittels HPLC. Mitt Klosterneuburg 38, 10–16
Kieckebusch W, Griem W, Lang K (1960) Die Verträglichkeit der *p*-Chlorbenzoesäure. Arzneim Forsch 10, 999–1001
Strahlmann B (1974) Entdeckungsgeschichte antimikrobieller Konservierungsstoffe für Lebensmittel. Mitt Geb Lebensmittelunters Hyg 65, 96–130

30.21 Furyl Furamide

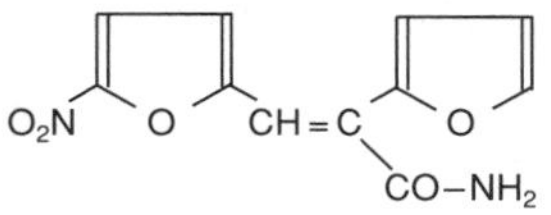

O_2N O CH=C O
CO–NH_2

Furyl furamide, *N*-(2-Furyl)-furan-2 carbamide and other nitrofuran derivatives have been, and are being, used to combat bacterial infections in man and animals. Because of their good antibacterial action, some of them used also to be employed on a small scale in food preservation, e.g. furyl furamide (AF-2), nitrofurazon (NZ-7), i.e. 5-nitro-2-furaldehyde semicarbazone, furazolidon (NF-180), i.e. 3-(5-nitrofurfurylidene amino) oxazolidin-2-on. The use of these products has been largely confined to countries in eastern Asia. Since it has become known that many nitrofuran derivatives are carcinogenic, they have been abandoned as food preservatives.

The LD_{50} of furyl furamide for rats after administration per os is 1.5 g, and for mice 0.5 g/kg body weight (Miyaji 1971 a). Additions of 0.2% furyl furamide to the feed of rats cause enlargement of the liver and histological changes in the liver within one week (Miyaji 1971 a). Changes are also apparent in the enzyme activity compared with that of control animals (Park et al. 1976).

In a 2-year experiment with rats, doses of 0.2% furyl furamide, relative to the feed, lead to increased mortality, whereas 0.0125% produces no pathological changes. Similar results were obtained with mice (Miyaji 1971 a). Furyl furamide has no teratogenic action or any other effect on the reproductive capacity (Miyaji 1971 b), though it is highly mutagenic, like many other nitrofuran derivatives (Tazima et al. 1975, Ebringer et al. 1982). It is also carcinogenic (Nomura 1975, Takayama and Kuwabara 1977).

Nitrofuran derivatives inhibit electron transfer in the cell, various redox processes and malatoxidation (Scott Foster and Russell 1971). In addition, they probably influence the cell membrane synthesis. Furyl furamide can be used to preserve weakly acid to neutral foods since its action is not affected by the pH value of the product to be preserved.

Most nitrofuran derivatives, including furyl furamide, are classed as preservatives with an antibacterial action. Bacilli, staphylococci, sarcina, vibrions and coliform bacteria are all effectively inhibited, as are *Proteus*, *Achromobacter* and *Flavobacterium* strains. The effective concentration is between 5 and 50 ppm (Matsuda 1966). 5-nitrofurylacrylic acid also inhibits yeasts (Farkas 1978).

As the action of furyl furamide is independent of the pH value, the product was used almost exclusively for foods impossible or difficult to preserve with other preservatives. These included fish preparations (applied concentration 15–20 mg/kg), meat (applied concentration 4–6 mg/kg), tofu (a soya preparation), bean paste and similar east Asian specialties (applied concentration 4–5 mg/kg) (Kennard 1976). For a limited period there were some countries in eastern Europe where nitrofuran derivatives were recommended for preventing secondary fermentation of wine (Farkas 1978). For toxicological reasons they also failed to establish themselves in these countries.

30.21.1
Literature

Ebringer L, Šubíkí J, Lahitová N, Trubăik S, Horváthová R, Siekle P, Krajčovič J (1982) Mutagenic effects of two nitrofuran food preservatives. Neoplasma 29, 675–684
Farkas J (1978) Les agents de stabilisation biologique des vins. Ann Technol Agric 27, 279–288
Kennard CHL (1976) Another food chemical, AF-2, alpha-2-furyl-5-nitro-2-furanacrylamide. Int Flavours Food Additives 7: 2, 59–60
Matsuda T (1966) Review on recent nitrofuran derivatives used as food preservatives. J Fermentation Technol 44, 495–508
Miyaji T (1971 a) Acute and chronic toxicity of furylfuramide in rats and mice. Tohoku J Exp Med 103, 331–369
Miyaji T (1971 b) Effect of furylfuramide on reproduction and malformation. Tohoku J Exp Med 103, 381–388
Nomura T (1975) Carcinogenicity of the food additive furylfuramide in foetal and young mice. Nature 258, 610–611
Park YR, Lee Y, Sung NE (1976) A study on the acute toxicity of AF-2. Han'guk Sikp'um Kwahakhoe Chi 8, 53–60
Scott Foster JH, Russell AD (1971) Antibacterial dyes and nitrofurans. II. Nitrofurans. In: Hugo WB: Inhibition and destruction of the microbial cell. London – New York Academic Press, p. 201–204Y, Kada T, Mukarami A (1975) Mutagenicity of nitrofuran derivatives, including furylfuramide, a food preservative. Mutat Res 32, 55–80
Takayama S, Kuwabara N (1977) The production of skeletal muscle atrophy and mammary tumors in rats by feeding 2-(2-furyl)-3-(5-nitro-2-furyl) acrylamide. Toxicol Lett 1, 11–16

30.22
Antibiotics

With the introduction of antibiotics to the therapy of human and animal diseases, interest in their use for food preservation became apparent at the end of the nineteen-forties. The main reason why antibiotics appeared interesting was that few of the harmless preservatives then in existence had an effective antibacterial action. Secondly, it was hoped that a number of antibiotics could be used in conjunction with heat treatment to shorten sterilization times, although in practice this expectation proved misplaced except to a small extent in the case of nisin. Finally, at first sight the antibiotics appeared useful for food preservation because they are many times more effective than conventional preservatives and it is therefore possible to make do with considerably smaller applied quantities, which are seldom organoleptically detectable.

Only two antibiotics have retained any importance, albeit relatively minor: nisin (see Chap. 27) and natamycin (see Chap. 28), the latter having a fungistatic action. Antibiotics are prohibited for food preservation in principle despite being fundamentally suitable from the technical viewpoint, since it is feared that their regular ingestion in food would lead to acquired resistance and hence undesirable influences on their therapeutic application and sometimes lifesaving action. The principle adopted is that substances employed therapeutically must not be used as food additives. This applies in particular to penicillin and streptomycin, which have never been widely used in food preservation. The danger in using antibiotics for food preservation is the possibility of damage to the intestinal flora. This applies primarily to antibiotics with an antibacterial action.

Some antibiotics, e.g. aureomycin and tetracyclins, were used for a limited period to preserve fresh fish and other sea food, the antibiotics being added in quantities of some 5 ppm to the ice with which the fish were refrigerated at sea. Fresh meat and fresh poultry have also been treated successfully with aqueous solutions of oxy- and chlorotetracyclin (Partmann 1957). The antibiotics subtilin and tylosin, which likewise have a purely antibacterial action, were used on a trial basis as a way to reduce the stringency of the sterilization conditions for canned foods. Tylosin was also used for preserving fish products, mainly in eastern Asia (Goldberg 1964, Shibasaki 1970, Suzuki 1970). These products are no longer of any commercial importance.

30.22.1 Literature

Goldberg HS (1964) Nonmedical uses of antibiotics. Adv Appl Microbiol 6, 91–117

Partmann W (1957) Antibiotica in der Lebensmittelkonservierung. Z Lebensm Unters Forsch 106, 210–227

Shibasaki I (1970) Antibacterial activity of Tylosin on Hiochi-bacteria. J Fermentation Technol 48, 110–115

Suzuki M, Okazaki M, Shibasaki I (1970) Mode of action of Tylosin (I). J Fermentation Technol 48, 525–532

30.23 Spices and Their Constituents

Many spices contain substances known to inhibit micoorganisms. The interest in using spices or their constituents as food preservatives is founded on the widely held but erroneous belief that any natural substance is less harmful to the health than a synthetic one. This ignores the fact that misgivings may exist regarding the toxicology of certain spice constituents (Hachitani et al. 1985). Allyl isothiocyanate is a case in point (see Sect. 30.6).

Spice ingredients with antimicrobial action include aldehydes, organic acids, phenols and essential oils. Other substances used to be grouped under the name phytoncides, as it was not known which individual constituent was effective. A slightly more detailed study has been made of tomatidine in tomatoes, humulone and lupulone in hops (Hass and Barsoumian 1944), allicin and alliin in garlic, onions and horseradish (Marth 1966). Interest has focused primarily on the antibacterial action of these substances (Beuchat 1976). Some other spice ingredients are effective against fungi (Forstreuter-Künstler and Ahlert 1984).

None of these substances has sufficient preservative action for practical use, as is clear from the fact that even foods containing such spices as normal seasoning are prone to spoilage. In addition many spices, especially in ground form, are so heavily contaminated with microbes that additional disinfection is required to prepare them for food use. If spice ingredients had sufficient antimicrobial action, the microbes would not have to be destroyed by irradiation or, formerly, ethylene oxide treatment.

Even if the antimicrobial action of spices were sufficient, more extensive use would not be possible because of their odor and taste. A "good" preservative is required to have little or no influence on the odor or taste of a food.

30.23.1
Literature

Beuchat LR (1976) Sensitivity of *Vibrio parahaemolyticus* to spices and organic acids. J Food Sci 41, 899–902
Fortstreuter-Künstler M, Ahlert B (1984) Hemmung von Mikroorganismen durch Gewürzinhaltsstoffe. Lebensmittelchem Gerichtl Chem 38, 143–145
Haas GJ, Barsoumian R (1994) Antimicrobial activity of hop resins. J Food Protect 57, 59–61
Hachitani N, Takisawa Y, Kawamura T, Tateno S, Sakabe H, Asanoma M, Noda M, Ishizaki M, Ishibashi T, Kuroda KL (1985) Acute toxicity of natural food additives and mutagenicity screening. Tokishikoroji Foramu 8, 91–105
Marth EH (1966) Antibiotics in foods – naturally occurring, developed and added. Residue Rev 12, 65–161

30.24
Enzymes

Besides lysozyme (see chapter 29) there are other enzymes which inhibit or kill microorganisms and could be considered as potential preservatives. Their use, like that of bacteriocins (see Sect. 30.25) is known as "biopreservation". Examples of such enzymes are conalbumin and avidin in eggs and lactoferrin and lactoperoxidase from cow's milk (Beuchat and Golden 1989). Conalbumin and lactoferrin bind iron, making it unavailable to microorganisms, and thereby inhibit the growth of bacteria. The presence of these enzymes in eggs may explain the fact that raw eggs have greater microbiological stability than boiled eggs, in which such enzymes are present but in an inactivated form. Lactoperoxidase acts with the thiocyanate present in animal tissues and the hydrogen peroxide formed from the catalase-positive microorganisms to create an antibacterial system (Reiter and Harnulv 1984). Myavert C is the name of a commercially available combination of lactoperoxidase, glucose oxidase, glucose and thiocyanate (Ashworth and Turton 1995). A similar system contains glucose oxidase and glucose (Dobbenie et al. 1995).

Apart from lysozyme, no enzyme food additives have yet acquired any commercial importance in preservation. Their action is very limited and vastly inferior to that of the established preservatives. In isolated form most of them are also too expensive.

30.24.1
Literature

Ashworth D, Turton K (1995) An approach for formulations removing preservatives from consumer products. Cosmet Toiletr Manufact Worldwide, 81–85
Beuchat LR, Golden DA (1989) Antimicrobials occurring naturally in foods. Food Technol 43, 134–142
Dobbenie D, Uyttendaele M, Delevere J (1995) Antibacterial activity of the glucose oxidase/glucose system in liquid whole egg. J Food Protect 58 273–279
Reiter B, Harnulv G (1984) Lactoperoxidase antimicrobial system: natural occurrence, biological functions and practical applications. J Food Protect 47, 724

30.25 Bacteriocins

Bacteriocins are proteins or peptides which are formed by bacteria and inactivate other bacteria. Like antibiotics, they are secondary metabolites with no known function for the bacterium producing them. The main difference between antibiotics and bacteriocins is that the latter are synthesized on ribosomes and are effective only against other closely related bacteria (Lücke and Geis 1992, Barnby-Smith 1992). Bacteriocins also include nisin which, because of its special importance in food preservation, has already been discussed in chapter 27. Nisin is the only bacteriocin to have been toxicologically tested, approved in various countries and introduced for food preservation.

In principle, bacteriocins can be used in food preservation in their natural form or as protective cultures which form bacteriocins in foods (see Sect. 30.26). The potential use of these products and that of certain enzymes (see Sect. 30.24) in food preservation is described by the term "biopreservation". The use of bacteriocins meets the demand for natural substances as food preservatives. Before using bacteriocins or bacterial cultures which release bacteriocins it is necessary to ensure that the bacteria involved do not also form toxins.

The most commonly discussed bacteriocins are those from lactic acid bacteria (Lewus et al. 1991, Okerere and Montville 1991, Hoover and Steenson 1993, Wang June Kim 1993), which include nisin. Since bacteriocins inhibit only bacteria closely related to those from which they have been formed, bacteriocins from gram-positive bacteria are not effective against gram-negative bacteria such as salmonella and campylobacter. Bacteriocins have no effect at all on yeasts and fungi.

As proteins and peptides, bacteriocins are sensitive to heat. They can therefore be used only in foods which do not undergo any heating process, although nisin is an exception to this rule. Bacteriocins are also sensitive to proteases in food or other similar microorganisms and are quickly destroyed by the proteases of the body.

Bacteriocins other than nisin are currently of no real technical importance in food preservation. Up to now discussion concerning their use has been merely theoretical and speculative.

30.25.1 Literature

Barnby-Smith FM (1992) Bacteriocins: applications in food preservation. Trends in Food Sci Technol 3, 133–137

Hoover DG, Steenson LR (1993) Bacteriocins of lactic acid bacteria. Academic Press, San Diego

Lewus CB, Kaiser A, Montville TJ (1991) Inhibition of food-borne bacterial pathogens by bacteriocins from lactic acid bacteria isolated from meats. Appl Environm Microbiol 57, 1683–1688

Lücke F-K, Geis A (1992) Bacteriocine. In: Dehne LI, Bögl KW: Die biologische Konservierung von Lebensmitteln. Ein Statusbericht. SozEp-Heft des Bundesgesundheitsamtes, p. 34–45

Okereke A, Montville TJ (1991) Bacteriocin inhibition of *Clostridium botulinum* spores by lactic acid bacteria. J Food Protect 54, 349–353, 356

Wang June Kim (1993) Bacteriocins of lactic acid bacteria: their potentials as food biopreservative (review). Food Rev Int 9, 299–313

30.26 Protective Cultures

Protective cultures are cultures of harmless microorganisms which are added to foods in order to inhibit the growth of pathogenic or other undesired microorganisms (Lücke 1992). Their use is closely linked with that of starter cultures and microorganism cultures intended for flavoring and coloring.

The action of protective cultures can be assessed in a similar way to that of bacteriocins. Some microorganisms produce secondary metabolites, which have an inhibitory effect on others. Of greatest commercial interest are microorganisms that form lactic acid, which itself has an antimicrobial action (see Sect. 30.13). Only non-toxin-forming microorganism cultures can be used in the food sector.

Microorganism cultures are used commercially for preservation mainly in milk, cheese and pickled vegetables. The use of lactic acid bacteria enables very small quantities of nitrite to be employed in bacon (see pages). For various reasons this procedure is not currently of any commercial importance.

30.26.1 Literature

Lücke F-K (1992) Schutzkulturen. In: Dehne LI, Bögl KW: Die biologische Konservierung von Lebensmitteln. Ein Statusbericht. SozEp-Heft 4 des Bundesgesundheitsamtes, p. 16–33

Chapter 31

Packaging and Coatings

31.1 General Aspects

A distinction has to be drawn in food preservation between packagings, coatings and liquids in which foods are immersed with or without a direct antimicrobial action. The first-named can prevent or at least restrict microbial spoilage by protecting food from infection or reinfection, or they can keep away the oxygen in the air that many spoilants require. One old technique for keeping butter fresh, formerly practised in the home, was to invert a bowl of firmly kneaded butter in water and store it in a cool place. Actually the value of excluding air or partially evacuating a pack is frequently exaggerated since a number of microorganisms (anaerobic microbes) grow even better when air is excluded than when it is present, whilst others, especially some mold varieties, can thrive in air containing only minute quantities of oxygen. Packaging in the absence of air is pointless unless the foodstuffs to be protected have previously been rendered largely or completely germ-free.

Packagings and coatings generally create only minor toxicological problems because the majority are not intended for consumption, nor are they indeed consumed in practice. Attention need be given only to any migration onto the food by substances contained in the packaging material. Usually, food law regulations covering packaging and wrapping materials confine themselves to this aspect.

A number of packaging and coating materials are themselves susceptible to microbial attack and need to be treated with preservatives. In principle the preservatives used for the purpose are those normally employed for food since it is not always possible to prevent traces of preservative from migrating onto the food the packaging contains. Sorbic acid is frequently used as a preservative because it retains its efficacy even in the high pH range.

31.2 Lime Water and Waterglass Solution

Lime water, a saturated aqueous solution of calcium hydroxide containing some 1.26 g $Ca(OH)_2$ per liter, and waterglass solution, a mixture of acid sodium silicates, used to be employed for the treatment of eggs by immersion. Despite its pH value of 10 to 12, lime water has no direct antimicrobial action nor does waterglass solution. The effect of lime water is rather due to the fact that the calcium hydroxide is converted on the egg shell into calcium carbonate, thereby sealing the pores of the sterile egg inside. This prevents spoilants from entering the egg and thus keeps it fresh. The preserving action of waterglass solution corresponds to that of lime water, as it involves converting the sodium silicates into insoluble silicic acid on the egg shell. The direct antimicrobial action of waterglass solutions is only slight. So far as flavor is concerned, eggs treated in waterglass are superior to those treated in lime water, in addition to which the white of waterglass-treated eggs continues to stiffen when whipped. Eggs treated in waterglass are also more brittle than fresh eggs and, like those treated in lime water, have a greater tendency to burst when boiled, owing to their blocked pores. Besides lime water another well known substance was Garantol, a solid product which can be dissolved in water and is based on calcium hydroxide with additions of iron, aluminum and magnesium salts.

As a result of improved refrigeration techniques neither preservative is now of any importance.

31.3 Mineral Oils and Fatty Oils

Microbial growth is impossible in anhydrous oils and mineral oils. Mineral oil used to be employed sometimes for the treatment of eggs, the mode of action corresponding to that of lime water and waterglass in that mineral oils prevent microorganisms from penetrating into the interior of the egg by blocking the pores in the shell.

Fatty oils are employed for preserving fish. Well known examples are presalted fish preserved in oil, especially coley, pollack, cod and others supplied as salmon substitute. Real salmon, too, is occasionally preserved in oil. Sardines in oil are not an example of oil's preservative action. Despite the oil these contain, their keeping properties are the result of a heating process.

31.4 Waxes and Plastic Coatings

The natural waxes most used are carnauba wax and beeswax. Other products employed include a number of synthetic waxes, such as polyethylene waxes, Gersthofen waxes, mineral (fossil) waxes such as montan wax and ozokerit, as well as paraffin wax. Natural and synthetic waxes are also applied in the form of emulsions or solutions. After the solvent dries, a uniform film is formed. Waxes and waxlike substances are used wherever the preservative action is a salient consideration, especially in the case of hard cheese and citrus fruits. For citrus fruit the Flavorseal process is of considerable importance; this involves coating the fruit with solutions of waxlike resins in low hydrocarbons (Charley 1959, Long and Leggo 1959).

Apart from plastic films, which are employed purely for packaging and thus fall outside the scope of this review, a number of plastics are also used as coating compounds. Mention may be made in particular of aqueous polyvinyl acetate dispersions for the treatment of cheese. Compared with waxes used for the same purpose, these have the advantage of allowing the escape of gases formed during the ripening of cheese.

Like oils, waxes and plastic coatings do not have a direct antimicrobial action. Their action involves inhibiting the growth of the microorganisms found on food surfaces by partially eliminating oxygen or reducing the water activity.

31.5 Antimicrobial Packagings and Coverings

A distinction should be drawn between the aforementioned packaging or coverings and coverings having a direct antimicrobial action on the foods contained within them. The latter contain, within their mass or on their surface, a preservative whose purpose is to migrate partly or completely into the food and there exercise its preserving action. Since this means the preservative will be consumed, it is fully subject to food law regulations. Hence, in the majority of countries little distinction is drawn between the direct addition of the preservative to food and its indirect application through the intermediary of the packaging.

Two types of preservative are used to manufacture packagings and coatings with a fungistatic action, namely those with high and those with low vapor pressure. The former type, of which a typical example is biphenyl, migrates out of the packaging material through the vapor phase onto the surface of the food. This produces a uniform effect, even on irregularly shaped goods, without the need for direct contact between packaging material and packed food. The volatility of the preservative may, however, be associated with a tendency to diffuse relatively rapidly into the interior of the packed food, which may sometimes be undesirable on physiological grounds, because of its effect on the flavor, or for other reasons (Lück 1962).

The second group of fungistatically active preservatives, namely those with low vapor pressure can exercise their action only by diffusing from the packaging material or coating onto the surface of the food. Hence, the food and packaging material have to be in close contact. With preservatives of this type there is likely to be a certain depot action at the interface between the packaging material and surface of the food, i. e. at the site where mold growth occurs. A typical preservative of this class is sorbic acid, together with its salts (Lück 1962).

The importance of packaging and coatings with an antimicrobial action has declined greatly because it is always more expensive to employ such products than to add preservatives direct to the food. Such products offer advantages only in special cases.

31.6 Literature

Charley VLS (1959) The prevention of microbiological spoilage in fresh fruit. J Sci Food Agric 10, 349–358
Long JK, Leggo D (1959) Waxing citrus fruits. Food Preserv Q 19, 32–37
Lück E (1962) Fungistatische Verpackungsmaterialien auf Basis Sorbinsäure und Calciumsorbat. Dtsch Lebensm Rundsch 50, 353–357

Subject Index

Printing: Saladruck, Berlin
Binding: Buchbinderei Lüderitz & Bauer, Berlin